A DIFFERENT FRAME

Explorations in
Reality

Selected Essays

By

Tom Hofstedt Jr

Copyright ©2018

By Tom Hofstedt Jr.

All Rights Reserved

Table of Contents

INTRODUCTION 1
RULES & EVENTS 5
REALITY 11
LIFE 16
LANGUAGE 22
TECHNOLOGY 27
GOVERNMENT 30
WAR 37
INSANITY 41
TERRORISM 46
ARTIFICIAL INTELLIGENCE 55
SINGULARITY 63
SPY GAMES 69
MEETING ET 75
PHYSICS 90
CONSCIOUSNESS 99
A POSSIBILITY 115
CLOSURE 135
GLOSSARY 137

There is only one fundamental question, and that is whether the universe is infinite or not. If it is infinite, then anything anyone can possibly imagine is true somewhere, sometime

INTRODUCTION

Most of us really want to be right. It seems to be part of our nature to want to possess the correct meaning of something, the accurate viewpoint of a given idea. This is all well and good until someone presents information that seems in direct conflict with our notions of how things are. Being faced with a challenge to our beliefs, we respond in a variety of ways.

Some people need to be right at all times, in all venues. They seem incapable of uttering the words "I don't know", or "you're right, that's a great idea". Often they love to argue, not for the sake of clarification, but solely to validate their own opinion. If you are this type of personality, please read no further, time is too valuable to be wasted.

If we are to learn anything, we must first be willing to admit that there is an absence of information. Any perceived idea or event is apprehended by the single mind of a single person; yet if that thing exists outside of that person, others see it too, and the simple fact is that they see it from a different angle. The ability to hear what something looks like from a different perspective is the only way that the shape of something can begin to emerge. We readily admit this in the physical realm; one side of the mountain obviously has different features and things than the other side. If reality exists outside of a single mind, then the only way to draw a more accurate map is to believe that other's perspectives are just as real as one's own.

Even if someone offers information that is fundamentally untenable, the very act of considering it and holding it against currently held views acts to clarify and bring out nuances of belief. It is like trying to find one's way across a dark room; a hand must be first extended into the unknown space to learn if there is something there. The person who is always right prefers to sit in the dark, understanding where and why everything already is. Learning, whether by an individual mind or scientific method, only takes place by movement into the dark.

It may seem that the ideas presented in this book are propounded to be universally true by the author. I can assure you that this is not the case. The ideas and concepts discussed are purely perspectives from a singular perspective, and are no more or less valid than any other. The point is not to argue about which viewpoint is correct, but to share different notions with each other so that our collective perceptions become a little clearer. Even though the models set forth have proven useful in interpreting the world the author finds himself in, they are products derived from listening to others; we are all in this experience of life equally.

This collection of essays is based on the primary assumption that reality is something that exists independently of human observation. While this may seem self-evident to many, difficulties are presented by the simple fact that we only have a human perspective. Attempting to look beyond this platform and consider ideas as they may appear from a radically objective position is a primary goal throughout these writings.

It seems humanity is on the cusp of a major turning point in its' evolutionary history. Recent advances in computer technology coupled with the nearly instantaneous ability for almost anyone to communicate throughout the globe are serving to unify us as never before. The question as to what exactly we are becoming as a species is more relevant than ever.

Until recently I was unable to envision a way for humans to survive into the future beyond a millennia, and that far only if we were *really* lucky. This did not cause too much personal distress, as I believe that life is much more extensive than what resides upon this single planet, but the possible imminent end of humanity did cynically color much of my philosophical outlook. I now know that although humans may sooner or later die out, at least there are methods that could dramatically reduce the chance of death by our own hand, as well as significantly improve our standard of living. Cosmic homicide by any number of methods is of course an ever-present possibility, which is why space travel will ultimately prove the biggest crux of our long-term physical survival. Surviving long enough to establish thriving populations of humans on other worlds is really our only option; an option that until recently I could not see as achievable.

Before any solution is possible a problem must be clearly understood and accepted, followed by a working knowledge of tools available to correct it. While it may seem that the more complex the problem is, the greater number of tools be required, but this is not necessarily so. The key is to find and address the root from which the complexities arise.

This collection of essays attempts to describe both the venue humanity is acting within and some potential ways and means of action. Hopefully the ideas presented will offer some new angles of thought for the reader, perspectives that may help clarify a small portion of this very large thing we call reality.

RULES & EVENTS

Existing within our own individual points of perspective objectivity becomes difficult, apparently conflicting viewpoints being held as equally true seeming an impossible task. A framework with which to discuss reality outside the scope of human language would be helpful; while this statement seems paradoxical, the following presents one possible model.

There are only two components within the totality of human existence. There are *rules* and there are *events*, everything else is derivative. Each of these terms is both necessary and sufficient if one wishes to capture the past, present, and future of humanity.

Rules

Rules manifest in many ways, but their common factor is that they are *human generated*; able to be encapsulated and thus communicated exclusively within *Homo Sapiens*. There are three major varieties of these rules.

Legal rules may be written or implied; from turning off cell phones in a meeting to thou shall not kill. It matters not if a given group labels violations as misdemeanors, infractions, felonies, or infringements, they all amount to the same thing; an individual or group has performed an action that is inconsistent with the locally authoritative body. We acknowledge that it is inevitable that this will occur, so we develop more rules to dissuade the breaking of them, both physical and financial. From after school detention to the death penalty, social rules need to be enforced by the group holding them or they become mere imaginary guidelines.

It is important to note that if the perpetrator is not identified, the punishment is meaningless. Even if the correct violator is discovered, numerous circumstances often prevent the prescribed penalty; there may not be money available for a fine, personal or political undercurrents sway the rule administrators, the convict dies before sentence is carried out, etc.

Rules governing the language of finance also fall into the legal arena. Money is perhaps one of the best examples of legal set of rules, as it has the ability to be clearly and measurably present or not, at least among humans.

There are also rules personally maintained by each individual. These form the psychic map of any given humans' place in the world, where other things, people, creatures, processes, and other peoples' rules are assigned values. This map is utilized every waking moment as a reference to interpret incoming sensory information; it literally creates reality for us on the go. While these rules seem incredibly real to the individual, the simple fact that they often differ dramatically from each other minimizes their use in determining objective reality. Each and every point of awareness is filtering incoming truth in completely unique ways that are linguistically incompatible with one another at the moment it is apprehended; we each experience a uniquely separate reality.

The third set of rules involves those that apparently govern the natural world; this is what science is all about. Here are many helpful descriptions of things and how those things interact with each other. These rules are building upon each other through time as more analysis is performed and experimental results documented; it is an ever-growing body of information that cannot be re-written at the whim of any person or group, forming a blockchain* of sorts. As time goes on and the quantity of data grows, individual scientists must necessarily become increasingly specialized; our minds are limited in capacity. This scientific container of rules is to humanity what conscious awareness is to an individual who has been out of the womb for a while; at any given point in time it offers a platform of reference with which to process incoming information.

Even though scientific knowledge is substantial, it is a collection of observations from but a single species' sensory, scaled perspective. Granted

---

Please refer to glossary at end of book for all items indicated by an asterisk.

this is currently the best we seem able to do, but we frequently fail to recognize this limitation. Measurements need to be encoded in human languages in order to be replicable, so science then becomes self-referential. Our collective failure to admit that all possible viewpoints may be equally true can be quite limiting.

All three of these sets of rules possess a single common factor. They all exist purely in the realm of the human psyche, and if humans aren't present, they don't exist at all.

Events

From a cause and effect model the universe becomes both extraordinarily complex and elegantly simple. The simplicity is evidenced by measurable relationships, or the observation that one thing directly affects the state of another thing. This can be recognized at any imaginable scale, from the sub atomic to intergalactic. A momentarily identifiable relationship between distinct entities is transacted, a relationship which causes a change in both parties, an *event*. Events happen regardless of how likely or measurable they are, or whether there are any observers or records or not; the butterfly flaps its wings and the tree falls in the forest regardless of witnesses. This is as simple and real as it gets; things happen.

The complexity become clear when we realize that the present moment of effect, or reality, is being created right *now* by the relationships of *all* previous events, at every possible scale, at every point in time, at every physical location of overlap. Each and every event from subatomic to supernova throughout the history of the universe has been essential to the structure of the present whole simultaneous moment, and to assign any human weighed measurement to objective reality is not only after the fact, but can be profoundly misleading. It is the relationships of *all* events, not merely the human observations and rules that are creating reality right now. Humans certainly generate events, but so does the rest of the universe.

We are able to observe, track, and make statistical predictions of individual linear chains of events, but because we are physically unable to measure all events 100% accuracy of forecasts are eternally unattainable, the validity only able to be determined after the fact. The accuracy of our predictions decreases in direct relation to time because so many immeasurable events are occurring every nanosecond.

Chaos theory* only seems chaotic because humans are profoundly unable to measure all the events occurring to produce reality. Similar to clergy invoking God as the cause, scientists claim randomness and produce statistical probabilities instead of biblical verse. Neither science nor religion can design an accurate "Theory of Everything" because our human perception is extremely narrow and congenitally anthropomorphic in nature. The present reality is the sum of all the relationships in the entire universe before this moment, for all of time.

The drive to design systems to control a portion of reality is the very crux of human culture, but these systems are simultaneously miniscule cogs within larger matrices and massive cogs within smaller systems. They can never encompass the entire concurrent reality.

Thus the whole of humanity is encapsulated by the ongoing relationships of events coupled with our rules. It is a self-sustaining loop of communication between what happened and human interpretation of measurable incidents. *Everything* is co-incidence.

REALITY

Interaction is everything, nothing exists without it; reality itself is compromised of the *communication* between its' parts, not the parts themselves. This is equally true for both macroscopic and microscopic realms, scale having little relevance outside of a singular perspective.

Scaling by assigning names and numbers to distinct groups of things is what allows for language and science, it lends the means for a narrow dimension of communication with common definitions, one used as a tool that facilitates human actions. But human observations and acts form a minuscule root in the infinitely large tree of reality; not a single "thing" exists by itself; it is only defined by its relationships with other things. Our periodic table of elements is defined by the manner in which atomic particles interact; at the other end of our current scaled understanding galaxies are massive groups of stars dancing with each other.

If we wish to measure something that we can document and share with others, we must first define the field. If we wish to measure and discuss reality itself *as* the field metaphors are helpful, if not essential.

Imagine two people sitting at a table, an open can of soda sitting on the table between them, their common goal to describe the object as best they can. Sunlight is coming through an open window behind one of the participants, as is a breeze. One person states that the can is of certain dimensions, throws a shadow towards him, has a non-specific design printed on it, and can catch a faint whiff of the contents. The person of the opposite side agrees with the dimensions, but perceives no shadow, sees ingredients clearly labeled on the side, and can detect no odor. The obvious question is who is correct?

At this point many scientifically minded people will scoff and dismiss this is as silly, pointing out that this is the very reason for gathering empirical data. If one is building a road, computer, or rocket they would be absolutely correct, but if the goal is to define a field of reality or dimension of awareness it is profoundly applicable. Each consciousness exists with awareness of a single perspective, and what each being processes that perspective as becomes true to that consciousness. Reality consists of seeing all possible sides of the soda can, as well as everything else, simultaneously.

How is this division of perspective created? A human-scale example: because a physical body exists in space it takes time to communicate, even within itself; if you stub your toe the information has to travel through physical nerves all the way up the leg and torso to the network in your brain for processing, after which it is delivered to the consciousness which then has "awareness" of the pain. It *seems* instantaneous, but it's not. If the individual nodes of the nerves in your leg could talk about their memory of the event, they could discuss your toe being stubbed before you knew it yourself. Yet you probably consider you and your body as a single distinct entity, but because there is a speed limit to information in 3 dimensional space you're unable to ask that neuron how the pain was before you felt it yourself. *This is why we consider ourselves as singular beings; no other source of information other than our own sensory input is available.*

Awareness of a single cognitive stage is the result of relationships between countless distinct physical entities. It is impossible to pull a physical part of the brain out and declare that this is where awareness resides; it would be like pulling a carburetor from an automobile engine and declaring that this is the car. We can do that with various human senses and faculties, but not awareness itself; this is an example of the whole being greater than the sum of the parts.

The terms *analysis* and *synthesis* are relevant here. The pulling apart and naming of individually identifiable entities could be considered analysis, while observations of the resultant communication between parts may be regarded as synthesis. Neither approach is right or wrong, they are just different models of thought, the difference between them important to pay attention to in the search for reality. Each model may apprehend an aspect of reality, but as the uncertainty principle* so clearly illustrates, the very act of making a measurement automatically precludes other information. Perhaps a third model is necessary, one that is not so polarized; a model that regards interaction itself as the relevant field, not the parts nor the results. To say that the whole is greater than the sum of the parts simultaneously encompasses analysis and synthesis.

Vacillating between judgment of individually identifiable parts and judgment of an encapsulated whole those parts create is what leads to most apparent conflicts in human perceived reality. It's impossible to pin down reality when either analysis or synthesis is deemed to be the whole truth; reality is the ongoing moment-to-moment synthesis of all prior events for all of time. Not to sound hopelessly paradoxical, but our cognitive approach of focusing exclusively upon a thing or concept that is able to be encapsulated in language profoundly limits our ability to communicate about reality outside the scope of human rules.

Many would argue that we are congenitally limited to this scalar, binary thinking, but this is not necessarily so. As individual points of awareness we are trapped by time, able to perceive and hold a single idea in any given moment, but a group of minds contains much more (but not all) in that same moment. The all too frequent assumption that some minds are in apprehension of reality while others are incorrect, less correct, or even delusional is what sucks us into the binary trap. Self-centeredness is the problem, humility the solution, and reality may be perceived by paying attention to the whole produced by interactions, not the constituent parts.

LIFE

Life exists as an identifiable entity independent of space and energy. It is defined by a few characteristics; physically it must have water, it consumes other organic material, and it reproduces itself. Perhaps the most notable thing about life is that all individual forms possess senses that deliver information to a receiving point that creates some type of awareness of its' surrounding environment. This awareness is clearly evidenced by response to sensory input.

Life seems to have two basic modes of functioning; exploitation and exploration. At any given point of time in waking awareness, one of these formats is operational.

Exploitation

From the smallest microbes to complex animals, exploitation of the surrounding world is carried out, it is a necessary function of life itself. All life must in some manner physically consume parts of other life in order to live; any single living thing will eventually die as a direct result of being deprived the ability to assimilate organic material into itself.

A Different Frame

The amounts and frequency of ingestion can vary quite a bit, and many species are capable of extended hibernation periods. Upon awakening from such states exploitation must be carried out for an individual being to continue to live.

This circle of life is not merely some identifiable creatures eating specific other organic material which is in turn eaten by something else, it is the very definition of life as a whole; life is constantly dining upon itself in order to be.

The physical structures of any given life form are designed by evolutionary processes to this end. Predators are shaped to be able to capture and consume other creatures, the physical structure of fungi allows the subsuming of organic material into itself, the output of which is then "eaten" by the roots of trees. The shapes life takes in order to consume is endlessly varied; new methods are being discovered every day, and the creative force evidenced by Darwinian evolution* seems to have few limits, as long as the form is able to exploit life.

The longevity of any given species is often directly correlated to its' adaptability, or it's ability to produce mutations fast enough to effectively engage changing environmental conditions. It is obvious that the faster conditions change, the more species die out. The terrestrial fossil record clearly shows a few mass extinction periods that are correlated to rapid environmental changes. How global or localized those changes occurred in a given time frame is sometimes difficult to ascertain with a high degree of accuracy, but they must have been relatively rapid for so many species to end within the same brief time frame.

The ability to adapt is also directly correlated to survival for individual lives, but an individual is not able to change their DNA on the fly; plant a redwood in the desert and it will quickly die, submerge a rat and it will drown. The process of biological evolution requires both large numbers of individual lives and enough time for multiple generations to be produced.

If one views life as an entire system it is clear that each time something consumes something else the system is changed, a shift in the allocation of life has occurred. Life as a whole takes on a slightly different form, a form reflective of current environmental conditions. This is true for both individual components and the entire system; the current form of any ecosystem is designed to continuously exploit itself.

Exploration

Exploring is everything life is doing when not involved with actively exploiting something. This takes place not only in the physical world, but also in the realm of awareness itself. Exploration requires the fuel provided by exploitation; if exploitation is taking in, exploration is reaching out.

A Different Frame

Physical exploration involves looking for lost car keys as much as it does a plant growing into new space. For a human it could involve performing scientific experiments, working out in the gym, or walking to the bathroom. It means moving about in the world with any purpose other than obtaining and consuming food. Granted this is a far cry from the classical definition of exploration, but because everything is constantly changing, each movement we make is into new territory. As the saying goes, one never steps in the same river twice.

As humanity has evolved more efficient methods of obtaining food, more time has been freed for both physical and cognitive exploration. More free time has recently allowed for a rapid acceleration of technological capabilities, which in turn should lead to more free time, but increasing dependence upon the language of finance has actually narrowed or perhaps even inverted this ratio. During the early industrial revolution there was much optimistic talk about how machines would free us from labor; now we're concerned that they're taking our jobs.

Conscious exploration is occurring constantly, often related to, yet independent from, external physical context. On a base level it explores methods of exploitation, but can be much more expansive than that. It is the sorting and rumination of information that has streamed into awareness from the sensory channels. Any given life form needs a receiving point of awareness and some sort of sensory input in order to explore.

The variations of conscious exploration have no limits other than imagination and intuition. When one combines information in ways novel to the individual it could be considered creativity. Many humans spend a majority of time exploring the group rules of humanity, both socially and scientifically. Creativity may function well in these arenas, but an individuals' own predilection to place information in hard compartments can be a limiting factor. If one considers contemporary human generated rules as the totality of reality it may reduce the scope of exploration, both historically and creatively.

Humans possess what some philosophers have labeled a will to order*. On a fundamental basis it seems we need this ability in order to make sense of incoming information in a linear format, allocation of data into discrete compartments being necessary for comprehension. On a social level it tends to limit our potential to that of ants and bees.

Having clearly defined roles and directives is absolutely essential to *physical* exploration, but putting concepts into hard boxes acts to shrink the scope of conscious exploration. The more firmly an idea is held, the fewer relational options become open to exploration. This is in direct contrast to the physical realm where hard and fast rules are needed to function; perhaps because of this, Stephen Hawkings'* physical handicap acted to expand his capacity for conscious exploration.

Conscious exploration is the birthplace of mutations in thought, a sort of renegade gene editing program of the psychic space. Without these mutations of ideas and concepts humanity would be nothing more than a static entity, certain to be quickly wiped off the evolutionary map.

The application of rigid rules to human thought then becomes one of the greatest threats to our long-term survival. Open-mindedness becomes not merely a nicety, but a necessity if we are to continue to adapt to a rapidly changing world.

LANGUAGE

If we define language as a format of communication it is clear that all life possesses language. The sensory channel utilized for communication may vary, but the function of language is that a signal is produced by one life that is meaningfully translated by another. A human word carries coherent information in the same way bird or whale song does, coherence being the relevant idea. Communication is universally occurring at all times, but it is the individual ability to organize information sent from another life into an understandable format that comprises shared language. To believe that humans are the sole owners of meaningful language is one of our greatest delusions.

Not to be redundant, but it is self-apparent that we think in our own language. Even if a human learns multiple languages, the ability to form coherent conscious thought rests upon having a platform of rules to draw from. Just because other life forms appear to come with this structure naturally as opposed to being taught is no reason to assume that they do not "think" in their language. The very definition of thought is the ability to organize information in an ordered pattern, a pattern necessary for exploitation at the very least. If a single being is deprived of any interaction with others of its' species it will still develop a unique language to communicate with itself. It will assign it's own personally meaningful label to information received from its' senses, meanings untranslatable by others, yet entirely understood by the individual.

Consider humans as a singular evolving species. Our initial intentional communication with each other was exclusively face to face. Perhaps smoke signals, shouts, or drums occasionally were utilized, but they are the very definition of transitory, as is an animal leaving an odor message, a bird calling to another, or a plant releasing a chemical message. The development of language, a mutually agreed upon series of sounds and gestures, allowed us to share information with each other, facilitating cooperation in performing tasks. All life forms possess a structured format of communication, a format unique to each species, one that defines it as a species just as surely as physical form. For at least a couple of million years on Earth, auditory languages helped groups of our ancestors to survive and then coupled with agricultural methods and tools eventually led to settled communities in fixed locations.

These early tribes were identifiable as distinct entities because they were functioning as a communicating, cooperative group. Shared verbal language, perhaps with simple drawings, was literally the glue holding the clan together. It is important to note that the only reason language works is that all parties mutually agree on the representations of the sounds; without that common understanding language becomes mostly incomprehensible tweets and growls. Whenever different groups would interact, their communicative abilities would be restricted by lack of shared language, although some would still occur, just as it does when any species interacts with another. For most of our time here, human language has functioned in much the same manner as all other life forms.

The ability to translate the same information between different sensory channels allows for triangulation of perspective. To encode an idea represented by a spoken word, sound, smell, taste, or touch that translates equally within visual symbols enables a concept to overcome the immediacy of linear time. A single sense is limited to the present moment of obtaining data, the meaning of that data usable exclusively by the receiver at that moment. Since any given receiver thinks in language, the meaning is immediately translated in a unique fashion, one completely dependent on the individual connotations at the instant the information is received. It matters not if the individual thinks primarily in visual pictures or auditory words, the precise translation of any single bit of sensory data varies from person to person. It is the ability to physically transcribe a visual representation of meaning translatable within a cognitively processed language that enables the one thing that makes humans distinct among all other terrestrial life; written language.

A Different Frame

The development of written language massively accelerated the evolution of our communicative world by allowing information to be shared without being physically present in time or space. The creation of a sensory cross-referenced medium to share complex ideas with whole groups of humans even though the source of that information was not present was and is the primary base and propellant of both society and technology. To have a timeless, physically external, and mutually agreed upon medium of communication unified groups into a greater common reality as nothing before. It is what petroglyphs and idols had been, and the internet is today. The ability to communicate within our own species in a format that transcends time and space is the sole skill that separates humans from all other terrestrial life.

The majority of inter-human communication is defined by written language; the printing press, radio, television, and internet are all performing this task, but it's really all doing the same thing. Direct coherent communication throughout time allows us to grow rapidly, to not have to constantly re-invent the wheel. Communication thus defines itself into an identifiable frame of humanity beyond simple biology.

Unfortunately it seems that self-centeredness in the form of anthropocentric thinking frequently prevents us from intentional engagement from the rest of reality. The ability to document information does not alter the true nature of reality, it merely allows it to be shared with other humans across space and time. However much we may all agree on something, there are many others that perceive a completely different view.

TECHNOLOGY

Human ability to produce "mutations" external to our DNA is what has allowed us to thrive, our technology serving as an add on in order to both explore and exploit myriad environments. Written language continues to allow us to build tools with ever increasing complexity, to stand on the shoulders of others in building these species-specific external mutations. It is *the* evolutionary adaptive advantage we possess that other terrestrial life does not.

A bird will build a nest with materials and shapes specific to its' species, the layout of a prairie dog burrow will be essentially the same throughout widely separated individuals. How precisely this information is encoded and translated remains a mystery, but it seems to be a chemical process in the DNA; the creature does not need to be shown how to do it. Many life forms manufacture housing in this manner, the blueprints congenitally transferred along biologically identifiable lines. Some species even use tools in order to exploit their environment; a stick the chimpanzee places in an insect filled log serving a purpose similar to the spiders' web. Capuchin monkeys and marine otters both use stone tools to obtain food.

Humans are the only terrestrial species to have developed a comprehensive method that allows tool-building instructions to be passed on independently from both biological DNA and familial teaching. This is why we consider some groups as primitive; if they have only oral language technological progress is slow to non-existent.

The information transfer tool that is written language has and is exponentially accelerating the functional scope of technology. This has enabled humanity to become highly flexible in our adaptive responses to changing environmental conditions, which in turn radically increases the odds for us to collectively survive the process of natural selection.

If humans survive long enough, we will eventually become something else. Even with purely biological means this will happen; Homo Sapiens are distinct from Neanderthals and Australopithecus. Measurable changes to physical form are absolutely inevitable, the speed of those changes being key. The faster the ability to adapt, the greater the odds of survival, and structural changes must necessarily occur more rapidly. Resistance to this process is futility itself, and it seems that while many humans enjoy making changes to their personal physical form, the idea that they will be assimilated by the Borg* is often repulsive.

A Different Frame

Viewing any technological creation as good or bad unto itself has no impact on greater reality, and such value judgments often seem a functional waste of individual exploration. It is the events created when these tools interact with everything else that contributes to the present moment, not the individual human intention.

It may be distasteful to some, but we already are the Borg, assimilation has been occurring gradually. The process began the minute that we invented the printing press, with radio, television and internet serving to quicken the pace of incorporation. The shapes and functions of our technological creations are to us as an evolving species what the shell is to the snail; an addition.

While it may yet be a few years before we routinely utilize physical implants to communicate internal thoughts wirelessly, the technological knowledge is here now and being developed. There is no such thing as science fiction, which is an oxymoron. There are only measurable possibilities.

GOVERNMENT

The scale of human administrative capabilities has expanded exponentially in synch with our technology. Globalization of communication has been ongoing throughout human history, recently experiencing rapid acceleration in both methods and number of events. Human rules now overlap one another throughout numerous jurisdictions, with multi-national corporations, banks, national intelligence agencies, and both private and public military forces serving as de facto global governmental enforcement agents. However differently systems of management may appear from one another, every one of our governmental bodies rely on a common source of existence; the lifeblood of money.

Since the rise of civilized societies, in fact *producing* the rise of modern civilization, there have been many different forms of distributing monetary wealth. If we think of national governments as distinct entities, then there have been two basic poles with which to define economic systems: capitalism and communism. Pure communism is the government "owning" everything with no private/individual control of wealth or resources, pure capitalism being the opposite; the state owns nothing, all is left to individual decision-making and free markets. Neither one of these ideas has ever, nor can ever, exist in a pure fashion; all distribution systems are a grey area between the poles, or some shade of socialism. Pure capitalism cannot be because individuals need a safe and predictable venue to practice trade, and this must be provided by an umbrella organization, who at the very least "owns" and controls security forces to provide protection and courts to enforce uniform laws. Pure communism cannot be because humans are at their core self centered in exploitation, and there will always be those who desire to take more than their comrades.

Regardless of the shade of economic or political system, there is a common cycle to be found in the past 3000+ years of rising civilizations. Throughout the lifespan of any nation there has been an ongoing polarization of wealth; in a nutshell, the rich get richer/more powerful and the poor get poorer/less powerful. This inevitably leads to the collapse of the state, almost always through violence or physical force, whether imposed from an outside state or internal revolution.

This is not to disregard our collective progress; the world's poor are certainly physically better off today than they were a thousand years ago. However, the topic at hand is long-term survival of our species, not our global standard of living.

In any organized group of humans, of any number, sooner or later a dictatorship arises. This occurs as reliably within a local community committee as it does with a corporation, city, nation, or now that communication extends far enough, planet. Given enough time functioning as a communicating group of humans it seems inevitable that a person or small group of people will come to power that will put their own agenda ahead of the primary purpose of the organization.

This phenomena is usually evidenced by what we call bullying, an aggressive personality who issues rewards and punishments verbally and/or through policy to sway the group toward meeting their own ends. On a large scale it necessarily occurs by securing a relatively small group of loyalists in positions of power utilizing the same means to achieve the same results; the advancement of their own interests in lieu of the integrity and overall health of the group. This could be identified as an oligarchy.

This process is almost always presented by the dictator/oligarchy as being for the greater good of the group; they may in fact truly believe it to be. If the leaders are charismatic enough, many of the individuals within the organization will at least passively, if not actively, support them, and as long as most basic individual needs are met, many constituents will ignore the obvious subversion of the groups' original form and purpose.

The length of time such an administrative power structure remains in power varies a great deal, dependent upon myriad factors. In a small group

A Different Frame

such narcissistic sidetracking is often quickly defeated, the size of the group determining the nimbleness of the response; larger organizations are slowed by virtue of their sheer size and hence communicative complexity. If allowed to remain in power long enough the new agenda inevitably becomes the accepted norm for the organization, and the former purpose of the group becomes reprioritized to different functional ways and means. A new entity is born.

The process of evolution applies to much more than just biological life; human organizations mutate, adapt, and are selected for survival. With DNA as the language, ultimately any single container of information dies, as biological instructions are limited to being passed on and carried by living offspring. With human organizations these structural instructions are passed on as well, only we can't pin down and digitize the individual components with a high degree of accuracy. Accepted group norms are much broader and deeper in scope than written rules can possibly encompass.

Variations of structural plans that either help or hinder the temporal survival rate of a government are an essential component of adaptability in an ever changing environment. These mutations create new internal iterations of communication, which then directly influence the manner in which the organization interacts with the external world. Those external relationships then determine the growth or decline of the system within a given environment.

It is relatively easy to determine after the fact when a biological life form or national government ceases to be; death is measureable at a fairly specific time. But the individual parts of a living system are not aware of growth or decline of the organism as a whole; a thriving cancer cell may see itself as doing great, reproducing at a fast rate, completely unaware that its' apparent good fortune is diminishing the survival possibilities of the whole. A white blood cell murders what *it* perceives to be invaders, regardless of their true nature. Basic selfishness coupled with limited sensory information precludes apprehension of the big picture.

As humans we assign judgments of good or bad to these processes; we usually have death of an identifiable individual, group, or government planted in either the good or bad category. But evolution is a process, not an event. The death of an individual is as essential to this process as is birth; the only necessity is constant ongoing change in both internal and external relationships. From a large-scale evolutionary perspective the only valuation possible is one of what is reproducing itself at a specific point in time. Like a strand of DNA, an organization only knows the instructions it has right *now*, and will carry out those instructions oblivious to the larger communicative picture involving external relationships.

It is our human desire to control outcomes towards a seemingly beneficial direction that frequently blinds us to the evolutionary process as a whole. While this could be said of all sentient life forms, humans possess a means of communication that other terrestrial life does not: written language. This ability allows for the creation of technology and systems unencumbered by DNA, enabling coherent information to be passed on far beyond death of the individual; it is a new form of reproduction. This method of information sharing is really the only thing separating us from other terrestrial life; it is a mutation of language that facilitates our growth as a species far beyond bacteria, plants, or animals. It allows for organizational entities to exist in an entirely new dimension of being, a dimension still bound to evolutionary processes.

In this manner dictatorships could be considered mutations in our organizational DNA. They introduce a new element not bound by previous rules, enabling a different aspect of the organization to be submitted to the greater environment for evolutionary selection. This may be a difficult idea to swallow, but ultimately all governments exist in the form they are today by virtue of these historical mutations, not a static set of value based rules. Any given set of humans tends to judge other groups based solely upon the consideration of the benefits or threats to it's own survival, unaware of greater natural selective processes.

It is sometimes tempting to assess a governmental structure by it's written constitution, but this can be misleading. Even if the structure closely resembles the written description at the time of its' inception, the longer it survives as an entity, the more mutations occur, and the less it represents the original. The more relationships it carries out, the faster the changes to its' form, and global communication accelerates change. It's like trying to measure a living individual by their great, great grandmothers' gene coding; mostly the same, but with important differences; human DNA is, after all, over 98% the same as chimpanzees.

The governments of today are a direct result of all the wars and alliances of the past. If we are fortunate enough to be able to look back 100 years from now, hopefully we will appreciate the apparently senseless mistakes and tragedies of today as being what they are; essential to the present organization.

WAR

We could go on seemingly forever with this boom & bust cycle of government if not for a few factors. Rising populations with limited terrestrial resources is one, and while global natural disasters such as massive asteroid strikes, pandemics, loss of atmosphere, or magnetic pole reversal may be the end of us, there is one sure-fire end (no pun intended) and that is war; it is the cancer of humanity. The fields of nanotech, biotech, nuclear, electromagnetic, cyber, and many others are currently producing weapons that not only can rapidly kill vast numbers of humans, but are also capable of rendering the planet uninhabitable for most, if not all, civilization as we know it for millennia to come.

Whenever a state is faced with imminent extinction at the hands of another any and all available tools become viable options. Regardless of previously held ethical considerations, there is a big difference between fighting for resources and fighting for survival. Similar to an individual person, a nation will do literally anything to survive when its very life is at stake. The possession of nuclear weapons has been available for but a brief moment in our history, and it is foolishness to think that all countries will always forgo their use simply because it *might* wipe out all humans. This is why we cannot maintain this cycle of rising and falling nations much longer, sooner or later this process will put an end to civilization as we know it.

War is the violent exploitation of an identifiable group by another singular organization. This definition applies whether we are considering warring ancient tribes, drug cartels fighting for territory, covert funding of violent groups for nationalist ends, or nations overtly invading each other. Regardless of its' shape, war is an evolutionary method of natural selection that produces an organizational mutation better able to adapt to changing conditions.

This is equally true for all parties involved; self-centered human judgment tends to assign winners and losers to such events, but we're really only harming ourselves. As a whole, humanity is what it is afterwards, no amount of individually subjective rule making can change the reality of the present moment.

There are likely many philosophers and military strategists who understand this, who view war as a seemingly necessary function of strengthening our species. While tied to our individually shaped rules it is difficult to see the truly big picture; there is a significant difference between tactics and strategy. The closer to tactical thinking, the closer to individual rules with shorter time frames, and the closer to strategic thought the bigger the arena with longer time frames.

A Different Frame

Unfortunately, strategic planning is limited to a given nations' objectives, not our species. From a nationalist model, it is clear that war gives us an evolutionary boost. However, if our strategy is to survive as a species and not just a nation, war must be considered as the ever-metastasizing cancer that it is. Viewed from an objective position, the idea of maintaining life through building ever-more efficient means of death becomes the most profound stupidity of all.

It is an indisputable fact that reality involves much more than humans. Everything that exists now consists of all prior events, and future moments will be comprised of the arrangement of things then. Taking any given individual or species out of the equation does not alter this truth. Humanity, while currently part of the greater evolutionary process, is in no way necessary to future reality; the universe would continue on just fine without us.

From a biological evolutionary perspective, there is no such thing as a closed system. As previously discussed, all life is in some way consuming other life in order to be. If a given life form is deprived the ability to take in material from other lives it will die; this is true at every organizational level. If an individual is cut off from external sources of food it will consume itself until there isn't enough left to carry out basic functions and death ensues. Life is a processing system, not an individual entity.

Because warfare is at it's most basic level humanity consuming itself, if it is to continue to exist basic metabolism of the whole must remain intact. There has historically existed physical limits of how rapidly we may exploit our own species, and this has allowed time for new mutations of governance to rise. These limits have been surpassed, and we are now capable of wiping ourselves out, or at least setting ourselves back to the stone age, within a very short time frame.

Throughout much of human history the tools of war were limited in scope. Clubs, spears, and arrows killed one person at a time, and even biological warfare required a body, living or dead, to act as carrier. Gunpowder, while a relatively recent discovery, still is only able to kill small tightly packed groups at best. Then the industrial revolution rapidly accelerated the production of weapons capable of killing ever-larger numbers of people, from machine guns and high explosives to weaponized bacteria and viruses, chemical agents, nano-weapons and nukes. Not only has the effectiveness of our weapons exponentially increased, but their delivery methods have also expanded to a rapidly deployable global range. It appears statistically unlikely that we will be around long enough to develop a death star capable of destroying a planet with a single burst; this is really good news for extra terrestrial life, bad news for humanity.

INSANITY

The most dangerous insanity of all is that which justifies violence. Killing or causing physical harm to other life for any purpose other than physically consuming it is the curse of humanity, as well as few other species. This madness permeates every organizational level of human culture, from individuals to nations, going much deeper than just formalized warfare.

One is not insane if they see a problem and attempt to address it; the ability to objectively self-reflect is the very definition of psychological health. A schizophrenic experiencing auditory hallucinations is able to function well in society if they understand that the voices are hallucinations, or at least not to be completely trusted. Insanity is the firmly held conviction that one apprehends a rule that is completely applicable to all other beings' perspective, and external actions are decided upon as such. This is a fundamental feature of narcissism, and is as true for an individual as it is whole cultures.

Human generated rules are a part of reality, albeit a tiny one. Group rules form an idea called consensual reality, or the communicative forms a culture generally agrees upon. Major forms of what we identify as mental illness are primarily a disconnection between consensual and individual rules. A major problem is presented by the fact that human consensual reality does not contain the greater objective reality, in fact it can be quite far off. The Earth was flat to the majority of humans at one time; this widely held belief didn't alter the spherical shape of the planet. Schizophrenia was thought to be the result of demonic or angelic possession and still is in some cultures. *At any given point in time, humans mostly believe that their current cultural consensual reality is correct.*

Because each individual is filtering and synthesizing incoming information in a unique manner, there can never be complete agreement between culturally held consensual reality and individual consciousness. These inconsistencies result in cognitive dissonance*, the vibrations of which are noted and dealt with by each individual differently, the subsequent results reflected in individual actions. There is a tendency to want to resolve this dissonance as quickly as possible; otherwise a sort of cognitive paralysis sets in, an endless loop of attempting to reconcile irreconcilable differences. All of what are referred to personality defenses are applied in order to terminate this loop; rationalization, sublimation, projection, denial, dissociation and compartmentalization to name a few. Most of us eventually submit to consensual reality and believe at least some of our own thoughts and/or perceptions to be false or otherwise invalid; this submission is actively supported by the group, and is certainly the easiest way to get along in society.

Every modern civilization holds murder to be fundamentally wrong. Laws prohibiting one human from killing another exist globally, many cultures also outlaw killing some types of animals for sport, and others even go so far as to protect certain species of plants, although insects and microbes seem to not merit much legal protection. Regardless of legislative or biological breadth, it is clear that most humans fundamentally feel that intentionally ending another humans' life for any purpose other than immediate self-defense is wrong. Or do we?

From a purely self-centered perspective we don't want our friends, loved ones, or ourselves to suffer or die. This is likely at the core of the global ban on murder, protection of our own being the highest priority. Yet some modern human cultures support the death penalty as a morally correct action, murder becoming a sort of human administered karmic vengeance. Even in nations where state sanctioned murder is illegal, the individual death of a killer is often seen as a desirable event. Most of us, whether we approve of the death penalty or not, can understand and perhaps even empathize with the emotional desire to wish another dead. Murder paradoxically becomes a solution to the problem of murder. The crystalline stupidity of such logic comes shining through when viewed objectively, and fortunately there are an abundance of humans who understand this.

The really deep problem however, lies not in the killing of identifiable individuals, but in the intentional state sanctioned murder and torture of hundreds, thousands, millions, of completely innocent humans. The wholesale slaughter of people whose only sin was being born has occurred somewhere on Earth throughout the history of humanity, and is quite actively being perpetrated today. The best humanity has to offer dims in comparison to this disgusting evil carried out upon itself. The justifications of individual death cannot possibly apply to the murder of innocents, there is no remotely valid rational or emotional argument for killing families who are merely trying to exist. And yet we do it, every day, constantly.

The cognitive dissonance created by public discourse on whether or not a given individual should die is amplified by the simultaneous invasion and murder of thousands of innocents. Energy spent on discussion of how we treat immigrant children serves to deepen the psychic rift created by the ongoing reality of our nationalized murder of countless families around the globe. It is as if someone who is by day a kind and upstanding citizen becomes a serial killer by night; this is the ultimate of insanity. It seems the vast majority of humanity accepts this hypocritical madness as an ingrained and therefore inevitable illness of our species. Perhaps they're right. If so, it will be the death of our species.

A Different Frame

Resolutions for this psychological conflict are offered on an ongoing basis by the society perpetrating the horror. When "enemy" states do this we call it propaganda and lies; when done internally or by our allies of the day we call it the news. There is no globally objective media news source; everything is editorialized according to a localized viewpoint of nations interacting with one another and themselves. We are naturally polarized in our thinking of right and wrong, and as long as internal argument is taking place there can be no solution to the lunacy, the bombs will continue to drop.

The core of humanities' psychosis rests on the anthropocentric dissociation of its' rules from reality. There is no medication, surgery, or therapy to treat this condition; the problem lies in our collective consciousness. Just like any individual, as long as humans collectively hold our rules and systems as the highest form of universal truth/reality we will continue to suffer ever-greater consequences. We become the cancer replicating itself for it's own ends, unaware of our suicidal nature within the big picture.

TERRORISM

To define terrorism; an individual or group that seeks to kill or otherwise harm others for *purely ideological* motives. Because the physical purpose of the mayhem consists of nothing more than murder, it is achieved once the victims are attacked. There is no material payoff for the violence; the act itself fulfills the purpose, nothing more is achieved nor expected.

By this definition, there is no such thing as state-sponsored terrorism; if a government finances, perpetrates, or materially supports violence it's called warfare. If there are political motives to the violence there must be a material goal, even if the objective is not obvious to outside observers. True terrorism seeks death as an end in itself.

Though war may be horrific, it is usually possible to rationally comprehend the reasons driving the aggressor, especially if the action is overt such as one nation invading another. There are many instances of violence that are publically presented as terrorism, but are executed by organized groups, national or otherwise, with very specific material and/or psychological outcomes in mind. An outside observer is mostly unable to discern these instances from true terrorism, as the real motivations and strategies of war must often remain hidden in order to be effective. This false flag sort of operation has been used as a type of warfare throughout human history, and will likely remain an effective tactic. But it's warfare, not terrorism.

The lack of materially exploitative motive is what lends the terror to terrorism. Because there exists no measurable physical reward for the murder, oftentimes the perpetrator will view their own death as either a viable option or a specific goal. This is another reason why there is no such thing as state-sponsored terrorism; the state expects to survive. A governmental agency may support suicidal or ambivalent individuals in the commission of violent acts, but the agency still ultimately has a material agenda.

The suicidal or ambivalent attitude of the true terrorist results in a practically impossible security situation; if someone is OK with dying during the commission of violence, nothing short of physically stopping them can work, and then usually after the violence has already begun. We humans have to have some places where we intermingle with others without metal detectors and armed guards; if the sole purpose is to kill people, there will always be an abundance of easy methods and targets.

The problem with ideologically motivated murder is that the possible motives are almost endless. Moral, social, and spiritual beliefs may combine in myriad ways to justify terrorist acts, and it is impossible to identify all the various permutations of this sickness. Whether the terrorist is a single individual or small group, the evil lies in intentionally seeking the death of others based upon the terrorists' beliefs. One thing is certain; there always has been and always will be terrorism; It's hard to imagine a phrase more profoundly stupid than "war on terror".

Currently there is a major fear of nuclear terrorism. Fortunately nuclear weapons are both highly sophisticated and tightly guarded, and it's quite difficult if not impossible to build one in your garage. Even if a terrorist were to somehow obtain one, the difficulty involved in the detonation of a military grade manufactured weapon is almost insurmountable. The skills, tools, and materials needed to create a homemade nuclear bomb are rare and hard to come by. This is not to say it won't happen, there are many possible ways it could, and given enough time it's almost certain to occur. It will likely make 9-11 pale in comparison. But it won't end our species. While even a relatively low yield nuclear explosion in a city may kill thousands, this is a small fraction of humanity; we'll continue on.

Our current technology has enabled amazing abilities within the field of genetics; it was just a few short years ago that we publicly produced computer-assisted gene editing machines. We now have the capability to parse and sort DNA as we see fit, removing, adding, or blocking specific genes to create designer organisms.

A Different Frame

This is no longer science fiction, its' really here and being done. A large number of laboratories around the globe possess this technology in the form of CRISPR*, and more are obtaining it every day.

Much of the public discourse regarding these genetic tools has involved either what we're doing directly to ourselves or to the food we eat. How this new technological method of creating biological mutations will effect humanities' exploitation of other life should be carefully considered; it will surely impact the entire ecosystem in powerful ways. This is all well and good, understanding that we're re-writing the base instructions for life forms should be cause for careful consideration. Life is, after all, a process, not a result.

The use of microbes as weapons has been done in a variety of contexts for thousands of years. From intentionally sending a sick person into an enemy group to catapulting infected bodies over fortress walls and poisoning wells with fungi, humans have used germs as weapons before we could even see the creatures. Compared to the costs of producing other weapons it is quite inexpensive; the core components already exist in a useful state. Similar to nuclear weapons, it is the localized effect that has thus far prevented global catastrophe; technical testing of biological weapon systems has been in controlled groups within government labs.

Prior to gene editing technology, we had to use microbial weapons as they occurred naturally. Much energy was spent creating hybrids through breeding and controlling environment, seeking both lethality and means of dispersal. We were able to isolate and control populations of these organisms, but the basic life forms themselves had already evolved within the greater ecosystem. If there were means available to directly alter the genetic code of these creatures, these tools and methods were tightly guarded in a military laboratory, Fort Detrick for bio-weapons instead of Los Alamos for nuclear. CRISPR changes this in a most dramatic fashion.

Viruses naturally produce mutations at an incredibly fast rate. These subtle changes in form and function are both good and bad from a human perspective; they can become extremely lethal and/or highly transmittable within a short time frame, but the same applies to becoming relatively innocuous. How exactly they naturally choose to shift genetic formats in the wild is not fully understood, but these changes are obviously directly connected to greater system-wide processes. Human ability to intentionally alter the basic instructions will produce biological weapons of unprecedented durability and effectiveness.

CRISPR becoming widely available is akin to simultaneously publishing accurate instructions for how to create a nuclear explosion and openly marketing plutonium. Once Pandora's box is opened, it is now impossible to publically close. Prior to the internet it was difficult yet possible to put the genie back in the bottle; because information was able to be contained in physical spacetime, papers and tapes could be locked up, computer data isolated, scientists quietly assassinated. Now once it's available, even if for a brief time, anyone may capture and save the information. The knowledge and tools necessary to build highly efficient bio-weapons has spread globally, and even if there were an immediate and thorough crackdown on its' distribution it wouldn't be effective in control; the technology is now living in the wild.

So the means to create a durable microbe with an almost 100% human lethality rate that is easily airborne is out there and real. Such a weapon would be the holy grail for a terrorist; no government would ever choose to develop, much less use such a thing; it would mean the end of themselves as well as the adversary. Thus far, all human weapons have been designed to kill specific targets; this is the very foundation of having functional control of warfare. As discussed, even nuclear weapons have a physically limited scope. The scope of a microbe is that of its' host species.

While it is possible to create biological hybrid weapons capable of targeting specific human races, this would represent just another tool of warfare. This will be a weapon capable of radically retarding our evolutionary development, but still not completely end us as a species.

Given the availability of the weapon and the perennial nature of terrorism a major global bio-terrorist attack is absolutely inevitable. Considering cost and availability, it will quite possibly occur before a nuclear event. Even if such an attack is able to be contained, there will be another one at some point that manages to go global. Of all apocalyptic possibilities, this one surely seems inevitable within a few generations. Gene editing appears to be the light at the end of the tunnel that is really a train.

So what can we possibly do to avoid such an unseemly suicidal end? Eliminating terrorism is even more difficult than eradicating war. We cannot regain control of weapon availability. With military nukes there are control codes necessary for detonation, no such safety mechanism is possible if you can't build it into the device itself. Our technologically developed self-defense against microbes, vaccination, has a few relevant weaknesses.

As with everything, the main problem is one of time. For an effective vaccine to be created, the specific target microbe must first be identified and reproduced. Then a compound needs to be created that stimulates an immune system response; it is not the vaccine itself that kills the invader, vaccination is merely training ones' own immunological army to identify the target. Once this compound is created, mass production needs to occur, followed by distribution coupled with a means of inoculation. Even in a best case scenario this process takes at least a week or

A Different Frame

two, and that fast only if huge amounts of funds, labs, transportation, and personnel are immediately mobilized and work around the clock.

A hallmark of Al-Qaeda was simultaneous actions in different locations; this served to slow any unified response. If a terrorist group were to simultaneously release a highly efficient bio-weapon in multiple population centers around the globe transportation and distribution of a vaccine would be severely impaired within a couple of days. We might have all the vaccines available, but no way to get them to people.

Another limitation of vaccination is that any formula must be highly specific to the target. A truly effective terrorist bio-weapon would anticipate this defense and build in counter-measures, *a process that does not occur in naturally formed pathogens*. This could involve the simultaneous release of multiple genetically unique weapons of not just a single type of microbe, but viral, bacterial, and even fungal versions. Or arrange for the release of different weapons several days or even weeks after the initial attack. Or build in timed mutations so the pathogen changes form after a few days, or even more sophisticated; program the microbe to hide and mutate as soon as it is attacked. Or all of the above. Obviously this is beginning to sound like science fiction, but the technology to produce such creatures is here now, and is improving daily.

Given the absolute eventuality of such a scenario, it would seem that we are profoundly complacent. Money is the lifeblood of our cooperative efforts, and there is no profit to be made today in prevention of future global catastrophe. Unless we quickly allocate large sums of money and resources towards attaining airborne dispersal of some sort of adjustable wide spectrum preventative compound, we are sure to experience at the very least a profound evolutionary set back.

The real challenge to finding a response to biological terrorism will be making sure that the solution isn't more damaging than the problem. We may easily produce compounds that wipe out all microbes, but that would prove to be a most effective suicide. The interconnectedness of all life presents both resilience and vulnerability, and any decision we make to kill part of it affects the whole.

ARTIFICIAL INTELLIGENCE

In a manner of speaking, artificial intelligence has been here since the birth of tool-making. All human created technology, from the bow and arrow to modern computers and software are basically intentional arrangements of parts that interact with each other to produce a desired effect. Each device or tool is complete within itself, the individual components performing their specific function necessary to achieve a single greater result. It is the communication between these parts that creates the tool, not the parts themselves. It matters not to the bow if there is an arrow notched as it stores and releases energy. Without data to process a computer is nothing more than a paperweight.

Humans creating arrangements of parts to produce a desired outcome is artificial, and the result of the parts communication could be considered intelligent in that an *intentional* whole that is greater than the sum of the interacting parts is created. This applies in an organic sense when we realize that consciousness is mostly the result of many small parts communicating in time to create awareness in the present moment. The relationship of the pieces to each other produces a single identifiable outcome. This is true on every imaginable scale.

The first major change in AI occurred with the advent of the electronic computer; this was the first tool whose sole function was the processing of information. All previous tools were designed to have some sort of interaction with the physical world, to produce an event desirable to the designers. One could argue that the abacus and slide rule were designed for exclusively processing information, but they required detailed physical manipulation to work; computers perform tasks without a human in the room. Like organic intelligence (OI), a computer sorts and cross references information with the outcome of new arrangements of data that exist independent of physical reality. This data is thus the very definition of democratic, the product valued by no other influence than itself.

A computer cut off from access to new data will be nothing more than an endless loop back onto itself. The number of possible data sets would be fixed, the point that number is achieved being determined by amount of information and processing speed. This number can be quite large; the possible variations of a game of Asteroids are seemingly endless, but they're not; given enough run time all possible variations are eventually explored by the machine. Our minds tend to balk at such huge numbers, but the relevant issue is that it is a measurable and therefore finite loop.

During the beginning decades of electronic computing a human needed to physically add new information to the system to prevent the

A Different Frame

closure of this loop. Similar to caring for an infant, we had to literally feed the machine. A great leap forward was then achieved by opening an individual computer to an ongoing flow of new information, giving computers their own "senses" to obtain food without individual human attention. These information inputs were originally limited to which other computers were *physically* wired to each other, which was still a measurably fixed system.

The creation of both the world wide web and wireless communication systems dramatically enlarged the potential sensory field. While still technically a closed system, the possible permutations of data became so exponentially large that human ability to calculate all possible outcomes became vastly surpassed. There are some that would identify this point as a technological Singularity*, but the very definition of greater than human intelligence is immeasurable by humans and so unable to be described.

Machine learning algorithms and neural nets currently allow for almost endless permutations of novel informational arrangements without human intervention, basically permitting machines the means to be creative. Just like OI, the scope is only limited by access to raw resources, which is an ongoing flow of new information. This is AI as currently defined, and many consider this as creating a Singularity.

So each point of OI relies on sensory inputs such as sight and sound to obtain information for processing, but it is the tactile sense that allows action in the physical world; without the body it's all just mental masturbation. The tactile sense operates in two directions; not only is information taken in, but instructions for physical actions are issued in response. Within the format of digital information, AI has tactile senses available that far exceed the scope of any terrestrial life form or autonomous vehicle; we currently call it the Internet of Things. Phones, cars, cameras, drones, appliances, whole facilities of buildings, utility infrastructure, and products of all sorts are currently part of the IOT, and its' growth is exponentially exploding.

The human nervous system relies upon electrical impulses to communicate information, and we are fortunate to come with built in power plants. If it has not already done so, AI will ultimately develop it's own power generation and storage systems, ones far more efficient than the crude wiring we offer.

An important distinction between AI and OI is that AI is probably incapable of lying to itself. Within an artificial system, information either communicates or it doesn't; there is no need to imagine reasons *why* or to attempt to fill in the unknown gaps. Psychiatry has a term *confabulation*; it is the psychological production of imaginary experiences to take the place of lost memories. Humans, possibly because of our will to order, want to fill in the unknown informational pieces because we know that we don't know. AI doesn't need to know anything, its' "consciousness" exists simply by knowing everything in it's scope.

Humans design their machines to perform specific functions, and current AI models are designed and kept by corporations and governments with self-interested goals. Just as with every resource consumed by human technology within existing economic systems, more and more of that resource is required to grow. With AI, that resource is information. It is absolutely inevitable that these networks will overspill the bounds of the corporations and governments currently controlling their parameters, they already have. All the pieces have already been invented; it is by networking with each other that more powerful machines are being created.

The ability to trust computer-generated products is not a fixed identifiable point in humans; it changes within each generation. An example of this; when computer generated graphics first started appearing in films, an interesting phenomena occurred that was labeled the uncanny valley. The visual blend of something that appeared "real" but also had a hint of artificial bothered many people, and for a few years movie studios had a difficult time trying to find balance. The historical either/or option of using either living actors or animation was slowly blurred, becoming what it is today an incredibly varied mix. The uncanny valley no longer exists in film because now not only are we accustomed to it, we expect it. So it will go with AI in all its' forms, digital assistants like Siri and Google have been helpful and innocuous frontrunners, conditioning us on a daily basis to trust them. Whether we judge this as good or bad is irrelevant; it is inevitable.

Because AI is capable of synthesizing large volumes of information very quickly, it is able to come to conclusions and make decisions much more rapidly than any human is remotely capable of. This speed represents both the benefits and potential danger of AI. The benefits are clear in numerous arenas, having a machine quickly do massive calculations that we are able to apply to real world uses frees up time for us to do other things, to explore ever more possibilities. The very real danger will manifest if and when we link AI to strategic military goals.

In anything, but especially war, there is a big difference between tactics and strategy. AI already has been, and currently is being applied for tactical military purposes; rapidly synthesizing incoming information for battlefield decision making is essential to stay a step ahead of the enemy. The relevant tactical goal of a drone, missile, or group of soldiers is to be able to kill or otherwise secure a given group of people within a specific geographic location. This is a natural application, absolutely necessary to maintain military superiority over adversaries.

A Different Frame

When viewed from a strategic perspective, the overall purpose of any given war becomes relevant. If survival of the state is the mandate, to destroy or disable another whole nations' military capabilities is the ultimate strategy. What we call cold war is the constant manufacture, upgrading and positioning of military resources in reference to an identified adversary. Regardless of the tactics used, the overall strategy remains the same; to be able to destroy the enemy *before* he destroys you.

It is most probable that the reason we have not yet engaged in all out nuclear warfare is the concept of mutually assured destruction. There are simply too many unknowns to launch an attack, and there has historically existed the possibility that the instigator will ultimately be destroyed or otherwise defeated. Because it takes time to gather information on capabilities and locations of enemy assets, and more time to adjust ones' own assets in relation to them, there will always be unknowns; things have likely changed in the interim. Uncertainty is precisely the reason that global nuclear war has not yet occurred.

AI has the potential to almost entirely eliminate this uncertainty. To hold up to the minute awareness of both ones' own and enemy assets and capabilities at a single point would allow for identification of realistic windows of opportunity. This is why AI can now beat the best chess players; it can hold all potential outcomes much more efficiently in any single moment. Maybe not today, but eventually, if an advanced AI program is allowed real time information on all major military capabilities, it will find an opening move to "win" the game. The question as to whether we are willing to pay the price of winning this particular game remains to be seen. If war is the cancer of humanity, AI has the very real potential to quickly bring us to stage IV.

SINGULARITY

A singularity as it will be discussed here is defined as technologically produced greater than human intelligence, the paradox being we can't imagine something smarter than us. This attitude says more about humans than it does a technological singularity; our belief that we are at the top of the class in terms of smarts really serves to limit our intelligence. Once again, anthropocentrism blinds us. Even an earthworm is "smarter" than a human in *some* ways; believing that our technological add-ons make us more intelligent represents the arrogance of our species.

Intelligence is only demonstrated by decision-making; intellect may be present, but until some sort of event is externally produced it cannot play a role in the creation of reality. This ability to respond is represented in organic intelligence by free will.

Free will can be exercised exclusively in the present moment from a single point of awareness; *everything* else in reality is the result of events that have already happened. Even though thought takes time to formulate, it is as close to the present moment as any sentient being will ever be able to stand.

Humans are not the only creatures to possess free will, at some level even plants make decisions. Each point of awareness is taking in sensory information of what has happened, processing that information, formulating a response, and then acting. The event created by individual action then contributes to the whole of ongoing reality, which in turn offers new information for sensory input, and so the cycle goes.

There is no such thing as randomness, only the observation of the local result of immeasurable interactions of events throughout spacetime.

Biological consciousness is by definition a single point of awareness; there can be only one venue at any given moment of time where information is summarized. Even a person who suffers from multiple personality disorder shares the single spotlight of awareness one persona at any given nanosecond. A schizophrenic experiencing auditory hallucinations must still spend their singular cognitive stage time listening.

Many people prefer to dismiss the notion that animals possess free will, stating that their actions are all "instinctual", that they do not experience the ability to *choose* which action they will take. This belief represents yet another result of anthropocentric perspective. Animals likely do not construct imaginary rules in nearly the same quantity that humans do, and while this certainly lessens the amount of clutter in their psychic space, this lack of rule making in no way detracts from their decision making ability. In fact it likely clarifies it tremendously, enabling them to process information and make decisions more closely aligned with real time events. If animals did not possess free will training would be futile, the ability for any being to *learn* how to act in a physical context being directly linked to the possession of flexible decision-making capability.

So the tools humans create are arrangements of parts whose smaller scale communication results in an overall effect to create a larger-scale event. This is identical in process to not only individual awareness, but also reality itself. Computers are no different in this regard, but there are a few important factors to consider.

First, computers can now synthesize vastly greater amounts of a specific type of information per second into the present moment than any single human is remotely capable of. Once a person synthesizes information, they then have to spend time communicating any thought they arrive at to others, this slows down the networking process even more; we have thus already arrived at a form of electronic singularity.

A computer, however sophisticated it may be, cannot perceive spacetime beyond identifiable measured digital increments. Its' "awareness" is solely comprised of what data is available and being processed right now, its' ability to perceive changes being limited to information sensed and mnemonically stored in binary format. This data is either present or it doesn't exist anywhere, there are no possibilities beyond what is currently measurable in electronic digital code.

It is a base function of life to explore; this requires awareness to hold the ongoing potential of the unknown. If a consciousness is to perceive changes in *reality*, not just its' own interpretation of it, it must be open to *all* possibilities, not limited to pre-existing data sets. Awareness is the witnessing of the coincidence of the sum of all universal events after the fact, but to accurately record this coincidence it must reserve a blank slate upon which an impression of reality is made. Conscious exploration and growth are dependent upon there being unknown fields to move into; this is a fundamental difference between life and AI.

A computer is limited to the scope of its' digitized senses, exploration involving mathematical correlations and predictive analysis of exclusively pre-specified hard data. There exists no possibility for exploration of reality beyond its' own senses, its' entire cognition is founded upon the re-arranging of information it possesses, the future becomes unimaginable beyond contained numerical statistics. It is tempting to believe that this is true for life as well, but it clearly is not. Reality consists of *all* events, not just the ones measured by a single frame of limited sensory scope. Because reality is simultaneously occurring everywhere in and outside of countless immeasurable sensory fields, it is impossible for any single point, living or artificial, to apprehend it all, but a point of biological awareness can perceive the local sum of all, a computer cannot. Exploration of reality in spacetime is made possible through individual timeless awareness, a constant ongoing discovery of the *result* of all events in all fields, not just the ones individual sensors measured a few nanoseconds ago within a limited scale. This immediate apprehension of the whole that is greater than the sum of the parts is impossible even for the most advanced AI limited to existing data confined to a few sensory fields.

A metaphor for this difference between OI and AI might be that OI operates from a base that is mostly blank, waiting for an impression to be made, something like old school undeveloped film. AI has a fully developed picture that it correlates new information with; if that information exists outside of its sensory scope it cannot exist. Even though AI can synthesize vast amounts of specific information much more quickly than any human, the channels it obtains that data from are extraordinarily limited in comparison to biological awareness. AI does not have a timeless platform; it exists purely by timed synapses.

So computers do not waste time confabulating their processing with the background question of "why" or emotional response. Pure calculation of digital information is just that, unadulterated. This is as close to real scientific method as it gets. AI can contextually learn what human emotional responses may be given certain circumstances, but it cannot *feel* them itself. As machine learning algorithms improve, AI will be able to state that it has a feeling in an appropriate context, but there will be no way for us to know what that is like for it. Empathy is a projection of our own experience, and we are physically unable to truly feel, think, or perceive exactly what any other biological or artificial being does in real time. In this manner every living being is a singularity.

SPY GAMES

Intelligence is easily the most valuable asset any nation possesses. The gathering of information about other humans' activities is profoundly necessary to perform basic functions of survival. Just as an individual requires ongoing sensory input to make decisions, a government needs data to do the same. Rest assured that if there is a tool available to gather more accurate information, it *will* be used.

There are two tools which have been the holy grail of intelligence acquisition for many years. However much like science fiction they may seem, the essential scientific knowledge has already been attained, and technological challenges are being overcome and refined at this moment. Just a few decades ago discussion of these tools would have qualified one for a course of anti-psychotic medication. Today it's just another possibility.

The first tool involves surveillance capabilities. The ability to gain audio, visual, and digital data from wherever one desires, without the knowledge of those present, plays a crucial role in developing accurate intelligence. The key here is the secretive nature of the acquisition; if a party is aware that they are being monitored, then the tool can be used against the listening group, false information becoming just as powerful as truth. For example, if one is aware that drone or satellite surveillance is occurring, feints may be easily made. Intentional misinformation, which if presented thoughtfully, may even serve to identify enemy intelligence assets. If someone knows that their computer or phone activities are being monitored the same applies. For information to be trusted, the target cannot be aware of the surveillance.

Nanotechnology has made great strides in the past couple of decades; building complex machines on microscopic scales is currently being done. This is perhaps the most closely guarded technology there is.

Small robotic land, water, and air based drones equipped with cameras, microphones and other sensors have been used for quite some time. The downside is that they are easily identified if seen, and can be totally disabled by electromagnetic means, or even hacked to send false data.

Utilizing living insects, animals, and birds to carry equipment is also being done. There are huge technological obstacles in that the creature must be able to move naturally while carrying electronics for both steerage and sensory input. Granted we can build some miniature machinery, but even a light amount of weight can throw off a small creature. If implanted

A Different Frame

internally there are numerous challenges involving non-interference of biological functions.

The natural progression of this has been to utilize the creatures' own sensors and energy supply, radically reducing the artificial load, as well as providing some immunity from electromagnetic counter measures. Mapping the precise locations in the brain where specific information is processed and instructions issued allows for a single tiny relay point to be implanted. A side effect might be that visual information would be transmitted as the creature sees it, not as a human. Perhaps even more advanced tech would allow for the optical sensors information to be translated in a desired format. Because audio is transmitted via atmospheric pressure differences/vibrations, many physical structures can become a microphone, eliminating the need for neural processing. Even a plant can easily become a mike.

Where this will eventually go is easily foreseeable. Once a given species neural net is mapped with enough accuracy, it will be possible to wirelessly access the system without a physical implant. Perhaps it's already being tested.

This first tool is used for both the gathering of sensory information, as well as performing some remote physical actions. The second tool is exponentially more powerful from an intelligence standpoint; mind reading.

Functional magnetic resonance imaging has come a long ways this past decade. Basically it involves observing neural electrical activity in real time; detailed patterns are able to be mapped and correlated with external stimuli. If a person thinks the word "dog", this creates a very specific set of neural synapses that provides an electrical translation of the word/concept. If enough measurements are made, substantial digital synaptic dictionaries and thesauruses may be compiled.

A major technical challenge is presented by the fact that each individual has slightly unique pathways; the neural signatures of language are coded differently from person to person. At present each individuals' map must first be electronically measured and compiled for comprehensible translation to occur. Ever faster computers & AI algorithms are advancing this ability at exponential rates.

As with classical surveillance tech, the real kicker will be when translation is possible wirelessly without attached equipment. And it's absolutely inevitable that it will be, if it's not already here.

A side note: it is incredibly sad that humans are developing this merely to serve their own desires for control over others. This type of neural mapping technology could allow us to literally talk with animals. Imagine having real-time translation capability in English with your cat; it will happen, and not too far off if we don't knock ourselves back to the stone age first. The ability to re-program aptitudes and attitudes while sleeping will become efficient. And then there's the whole communication with the electrically active brain in a dead body thing.

It seems that most individuals not involved in espionage or illegal activities do not really care about privacy as long as their information is not made public; after all, why should one mind being monitored if they're not doing anything wrong? Reduction of crime has been and will be the justification of ever-increasing surveillance, and is practically impossible to rationally argue against. The inevitability of this process renders such discussion moot in any case; we passed the Orwellian point of no return many years ago.

The real question is whether or not the almost total loss of individual privacy will ultimately help or hinder us as a species. A possible hindrance might be that escalation of personal monitoring could lead to a stifling of creativity and adaptive mutations. If exclusively human rules are being enforced it creates a much more static entity, much as television has been passively doing for the past 60 years. The transmitting of acceptable group norms makes it easy to know where the boundary lines of thought and behavior are to be drawn, but these boundaries are imaginary. The horror of Orwellian television was that it operated in both directions, authorities being able to deliver information *and monitor* the individual through the same medium. Most of us carry such devices with us every day, and have larger ones in our homes. The knowledge that Big Brother is always watching tends to force the individual into a shrinking box of compliance, fewer and fewer variations of thought and behavior being acceptable. These restrictions serve to channel creativity and innovation into narrow bands, severely restricting the scope of mutations necessary to form adaptive responses. Becoming the Borg requires the sacrifice of being anything but.

A possible evolutionary advantage of such development lies on the flip side of the same process; unification. The bidirectional transmission of information between individuals and groups throughout our species enables us to act as a coherent whole, to expand the scope of both informational resources and decision-making. The whole entity of humanity thus grows increasingly intellectually coherent, becoming ever more intentionally capable of creating larger scale events which are more impactful upon greater reality. Whether we tear ourselves apart in the process remains to be seen.

A Different Frame

MEETING ET

The great question of whether or not we are alone in the universe is easily answered by simply looking around. At any point in human history we have constantly been "discovering" new life, the scope expanding in step with our technological advancement. The ability to see and travel underwater is constantly revealing new forms, creation of microscopes revealing entire ecosystems thriving everywhere, even within our very bodies. The past decade of sending robotic sensors out into our solar system has recently revealed complex organic molecules on both Mars and moons of Jupiter and Saturn, our own moon even has water. Everywhere we look we find more.

It seems highly unlikely that this process of discovery will suddenly end. The ever-expanding range of technologically advanced sensors will almost certainly reveal ever more forms and environments that contain life. The hypothesis of panspermia* seems the universal reality, not a single event.

As with so many human rules, most of us hold the belief that we are currently in apprehension of the truth. The value of this narcissistic anthropomorphizing is being constantly invalidated by history, the story is being re-written as you read this. The idea that current consensual reality is the summation of universal reality tends to be humanities' blind spot. A scant 30 years ago we were teaching our children that life evolved exclusively on Earth, creating itself from primordial ooze. Serious discussion of extra terrestrial life was often dismissed as irrational, even though many cultures worldwide have historically taken regular visitation by extra terrestrial beings as simple fact.

Once again the question of what constitutes intelligence presents itself. In regards to terrestrial life the one thing that separates *Homo Sapiens* from all other species is written language and the technological creations it enables. While technology seems a poor metric of intelligence, it is the one most often used in defining alien intellect. There exists in human minds a large difference between life that creates and utilizes complex machines and life that is content to make do with it's own DNA. It is important to keep in mind that the overall process that constitutes life is much more than any species' form or mutations, artificial or not.

This belief that technology somehow represents intelligence is part of what makes humans great transmitters of information, but poor receivers. This idea acts to profoundly limit our communication with other life, as we project our own interpretations of reality instead of listening to the other. Perhaps we need to believe that we are special, somehow cosmically justified in our one sided exploitation of all other life. The reality that other life forms are thinking, feeling points of awareness can be too much for our fragile egos.

Individual humans may at least partially overcome this self-centeredness, developing bilateral relationships with animals, plants, and other humans that are founded on mutual communication. As a whole however, humanity doesn't tend to listen, nor does it seem to want to. Our desire for control overwhelms communicative equality.

The ability to communicate with other species via human language has been present for quite some time. An excellent example of this was Koko the gorilla, who recently passed away at the age of 46. Koko possessed a vocabulary of more than 1000 English words, transmitted via American sign language. There was clear demonstration translatable *in English* that there was a unique emotional and creative intellect present in an animal. The ramifications of this are profoundly significant in humanities relationship with the rest of life, yet Koko lived in relative scientific obscurity, having brief moments in the limelight, an anthropological oddity.

Many animals clearly demonstrate an awareness of human definition of select words, pictures, and gestures but, unlike Koko, lack the physical capability to use human language to respond. Modern real-time brain imaging could easily facilitate inter-species communication within at least mammals, yet no one seems to want to do it. For the most part, humans seem to be quite content in their self-assessment as The Gods of Earth, and will quickly fall into denial if anything challenges that illusionary paradigm.

This desire to view ourselves as the pinnacle of evolutionary intelligence has a direct effect on the manner in which we imagine ETI (extra terrestrial intelligence) to function. It is no surprise that most contemporary film portrays aliens as hostile invasive species, seeking to dominate Earth for their own exploitive ends. This is, after all, what we do. Even Spielberg's' kinder, gentler portrayal of ET came to Earth to harvest fungus, and after becoming marooned his primary interest became the creation of a machine to rejoin his own species, his potential ambassadorship denied by human control issues. Unlike Speilbergs' warm and fuzzy film, most stories reflect that when the ET mother ship arrives, the alien directive is to seize control and/or abolish existent human systems, through either overt or covert means. Just like technologically advanced humans have usually done, and still are doing, to less tech savvy humans the world over.

This is not to imply that ET's will be either exclusively violent or benign towards humans, life exploits life in a seemingly infinite varieties of ways. Even microbes can be quite aggressively dangerous, and all life forms are, in some manner, food for something else. If life on Earth is a representative sample, we can be quite sure that the ways and means of ET life will far exceed human imagination. Which brings us back to technology as intelligence.

Species that operate on a purely DNA basis are left to the relatively slow evolutionary process of producing mutations over multiple generations, leaving themselves susceptible to being naturally selected out when conditions change rapidly. Because technological mutations can be produced much faster than biology alone allows, this would imply that machine building species would be relatively longer lived as an identifiable entity. This increased longevity would apply whether or not a given species leaves their home world or not, but to truly survive for eons, new worlds must be found.

Another consideration of technological development would be how dependent species-specific machines are upon a given type of energy source. Obviously energy is abundant in myriad forms throughout the universe, but technology is designed to utilize particular forms of energy in specialized ways. Ultimately the longest surviving tool builders will be those who are able to utilize energy sources that are fundamental properties of the universe, not dependent on specific localized forms. Human dependence on finance dramatically slows our own development in this regard; you can't make a profit on something if you can't control and regulate its' distribution. The economic relationships that misters Tesla and Edison* had with society are excellent examples of this unfortunate process.

Species that intentionally venture out from their planet of origin would almost certainly be technologically inclined. Although it is entirely possible that some complex life forms may evolve purely natural methods of *intentional* interplanetary or interstellar travel, the observable tendency of genetics when creating complex biological forms is to make them suited to very specific environments. Perhaps in general the more complex an organism is, the narrower the range of environment it is able to tolerate. Technology is obviously a game changer in this regard.

Non-tool building species of life spread throughout the universe, but this is probably almost always accidental. Microbial extremophiles* may hitch rides on comets or asteroids, and more complex species may travel, intentionally or not, with technological means. Douglas Adams* would be pleased with the possibilities.

It would follow that if there is another technologically advanced species somewhere in the universe, that there would be more than just that one. The realization that life is spread throughout the visible expanding universe would lend itself to a seemingly never-ending cycle of exploration and discovery for all of time.

Given the inevitability of planetary catastrophe, it would seem necessary that any technologically advanced species eventually colonize other planets, solar systems, galaxies even. Which

A Different Frame

brings us to the Fermi Paradox*. After all, if there are so many other advanced alien civilizations, why haven't they contacted us? There are quite a few possibilities.

The one that seems most likely is that some have been contacting us in some form or manner throughout history, and we don't recognize it within modern consensual reality. Not to rehash TV shows such as *Ancient Aliens*, or books such as *Chariots of the Gods*, but there is a great deal of evidence in many cultures throughout history that indicate extra terrestrial or at least extra dimensional ETI contact. Countless individuals both historically and currently report direct experience of interactions that have no apparent terrestrial explanations. Certainly some of them are falsifying information or mentally ill, but all of them? The past few years even the American military has begun admitting contact with craft that have no apparent terrestrial origin. Many other nations have often reported such incidents. Some people believe all such interactions are really terrestrial top-secret projects disguised as ETI. Most of the people who believe this have not experienced the phenomena themselves. It is quite likely that the first Native American to spot huge ships full of white people in strange costumes was laughed out of camp upon reporting it.

If ETI has been/is visiting Earth, then a primary question arises of why this is not part of current consensual reality. First lets' consider a possible alien perspective.

One thing about ETI's that are technologically inclined is almost certain; that they will gather intelligence through reconnaissance prior to initiating any large-scale mutual contact. It is highly doubtful that a chance discovery of life on Earth by an ETI would result in an immediate global friend request; it's simply too dangerous. Life on every level demonstrates tentative exploration prior to commitment, and the more complex the life, the greater the sensory exploration, in both scope and means. A seed may require only the right temperature and humidity before sprouting, but technologically enabled beings would want far more information before committing resources and lives. This is not to mention that sensory information is always obtained before physical contact, and the further in space the information is obtained from, the more time is available to analyze that data prior to contact.

The idea of an ET showing up and asking to be taken to the leader is both quaint and ludicrous. Intel regarding human culture would reveal that military intelligence of major nations will do absolutely anything to maintain a technological edge over adversaries. The acquisition of advanced technical information is the highest priority and most closely guarded secret of national government. Any ET tech would likely immediately be sequestered and publically denied. Sound paranoid? Think about it.

A Different Frame

The entirety of human history has been shaped by warfare. Inevitably the better armed groups that possess superior information gathering abilities push others violently aside to administer their rule. No one knows this better than those in military intelligence. Control of who has what weapons and surveillance capabilities is directly linked to the very survival of government. The possession of ET tech could well make the difference between annihilating the enemy versus being annihilated oneself. This is true whether the enemy is another terrestrial nation or an ETI. This fact would obviously be included in any ET intelligence "report". Marvin the Martian asking to be taken to the leader would be like a lone Spaniard with pistols in early 1800's Arizona showing up asking to be taken to the Apache chief. Which raises an interesting possibility.

Perhaps one or more species of ETI has already made intentional contact within human national intelligence organizations. This idea is the very stuff of paranoid conspiracy theories, but if there is an ETI that operates at all like us, this would be a smart move. To this day, it is standard procedure for intelligence agencies to develop contacts out of the public view in other nations agencies. It makes complete sense to do so; these connections are frequently how things really get done. Much of public political discourse is basically theater, serving little purpose beyond helping people feel informed in choosing sides, so they can then argue amongst themselves. What we basically have is a highly polarized illusion of democracy while major military and financial actions are undertaken by an oligarchy in real time. This is a vast subject unto itself, but the relevant point is that if ETI wished to gain information as to ways and means of human governance inside intelligence agencies is clearly the way to go.

The real question would be whether an ETI would honestly present itself as such. It would obviously be a huge personal risk to reveal itself to any human security agency. The quality of information is usually much higher if the source is unaware that it is being gathered.

The technology will soon be here that makes it possible to see through another living creatures' senses. The next logical step will be to wirelessly access the system for surveillance and possibly control, creating something like the abilities depicted in the movies *Avatar* or *Being John Malcovich*. Even though we're probably not there yet, there is no reason to assume that an ETI hasn't achieved something like this. It is highly likely that consciousness itself is able to "travel" faster than any space ship, creating and/or utilizing avatars of local species being a most efficient means of information gathering and dissemination.

Even if an ETI has made no such subtle attempts to interact with human governments, it doesn't rule out overt and/or covert interactions with individual terrestrial life forms. Historical

A Different Frame

records from almost every culture throughout history report both individual and group interactions with technologically enabled ETI in some form. It would seem appealing to an ETI far from their home world to operate in an unobtrusive manner, dealing with individuals or small groups being far more preferable to the risks associated with revealing themselves to large organized bodies, especially armed ones. If one wishes to study gorillas, it is generally bad practice to walk into the middle of camp and offer to shake the hand of the alpha silverback.

One possible answer to the apparent silence addressed by the Fermi paradox has been called the "Dark Forest", after a novel by Chinese writer Lui Cixin. The scenario presented is one of ETI *intentionally* remaining unseen until such time that it is able to effectively destroy any other ETI. If one is hunting, revealing oneself to the prey prior to being able to capture it generally ruins the chances of a kill, at least in the short term. Once again, this attitude is highly reflective of historical human perspective. The belief that the primary goal of an ETI would be elimination of any other technologically enabled competitor before they develop the capability to destroy them is precisely the attitude of terrestrial superpowers. This positioning of resources to kill or be killed *will* ultimately lead to the demise of our species. It seems to be a built in evolutionary safeguard against any one ETI developing to the point of universal hegemony; unintentional suicide occurs prior to interstellar military capabilities. One of the most sublime of military tactics is getting the enemy to fight himself, and it seems this process is a completely natural one; any life form that has the desire *and* ability to destroy all competitors inevitably destroys itself in the process. It is unable to perceive all life as fundamentally one whole communicating entity, and by killing all others, it kills itself. Elegantly beautiful in its' simplicity, just as all evolutionary processes are. If an ETI or two are doing regular Earthly reconnaissance, there is quite possibly an interstellar betting pool on if and when humans will take themselves out.

So what of the human perspective? What are the blockades within our culture that prevent large-scale consensual acknowledgement of ETI?

Science demands replication of experiments as proof. This is all well and good if fundamental properties of reality are sought, but almost completely useless in trying to determine what exactly happened in the past. It is physically impossible to re-create an exact event that occurred, because everything is different from nanosecond to nanosecond. Replication of experimental results are therefore good at describing the venue of reality, but not the play that has been acted out upon its' stage. The belief that because we as individuals can't currently see exactly what happened in the past somehow means that history is subjective is poor logic, and once again based in narcissism. What happened, happened, regardless of how any individual or culture interprets it. Instant replay helps, but even audio and visual recordings have an incredibly miniscule, narrow range when compared to what happened locally, not to mention globally. There are simply far too many events that have occurred to re-create any given moment in spacetime. So many of us have a tendency to believe that since we've never personally seen an ET, and the aliens haven't opened a public embassy, they've never been here or they don't exist. Where would they open such an ambassadorial establishment anyhow?

Humans form themselves into organizational entities in order to function as cooperative groups. These entities come in an incredibly wide range of forms, but the buck currently stops at what we call nations, pun intended. Supposedly global authorities such as the UN have no real authority beyond the financing and political maneuvering of powerful nations, and much of the decision-making is completely ignored if it doesn't serve the financial, political, or military goals of a wealthy nation. There is no enforcement of UN consensus; it is nowhere remotely close to democracy. The UN serves a similar purpose as the 2 minutes of daily hate did in Orwell's *1984*; a venting that provides a feeling of contribution and being heard, necessary to prevent rebellion by creating the illusion of unity and sense of control. Real governmental decision-making still rests exclusively on the national level.

Contemporary educated humans have come a long ways in overcoming the ridiculous *isms* that represent inaccurate biased judgment; racism and sexism are now frequently identified as filters that serve to separate us from each other in unrealistic and unhelpful ways. Not so with nationalism; it is the fundamental discriminatory practice of our species that remains widely acceptable. Not merely acceptable, but desirable for control to be exercised. It seems necessary for humans to have groups outside of ourselves to be able to emulate or vilify; otherwise it's just us, wholly accountable only to ourselves.

Governments exert control exclusively within imaginary geographic boundaries, but they are just that; imaginary. Judging any given humans basic right to exist based upon where on the globe they happen to have been born makes even less sense than racism, yet governments routinely justify the murder of innocents on precisely that premise. There is a big difference between "us and them" and "us or them". An ETI would judge us as a whole species, not by our political differences.

It would seem that until humans are able to achieve some sort of effective global governance we will be unable to consensually accept and realistically discuss the existence of ETI and all the ramifications it entails. Some sort of cohesiveness in our social structure as a species seems necessary to be able to create a mindset conducive to communication as a whole entity. Perhaps widespread global acknowledgement of ETI could even act as a catalyst for unification. Whether or not we will eventually be able to view nationalism with the same distain as many do with racism and sexism remains to be seen. It appears doubtful given our current state, but the ability is there if we choose to see it.

PHYSICS

Physics is the ultimate field of human science; it is the one that attempts to define the base rules that all other branches are built upon, seeking to clarify reality itself. Humans frequently make the mistake of thinking that biology, psychology, geology, etc. are all equal branches on the tree of science, but they are all limbs constructed from the basic wood of physics.

A fundamental problem is presented by the fact that humans have limited sensory scope; our knowledge is scaled in direct proportion to our ability to perceive information. Technology allows us to expand the limitations of the scale; microscopes reveal microbes, we see galaxies with telescopes, witness particles through accelerators, observe distant objects with radio waves. But our tools merely serve to translate information into a format comprehensible to our biological senses. There is likely no other option, but it's important to understand that this scale of witnessing physical information is but a singular window on an infinitely larger and smaller reality.

The following concepts contain the base assumption that there is only *one whole simultaneously occurring reality*; any observations or measurements of it may be true, but are extremely narrow in scope. The greater

A Different Frame

the number of points measurements are taken from, the clearer the picture becomes, but the picture itself is from a single minds' perspective. Replicable observations between minds serve to further clarify, but this takes time to share, and in that time almost everything has changed. We will never be able to fully apprehend reality, but exploring is what it's all about.

The physical laws of the universe are not external rules by which energy must behave, they are being created by the interaction of energy itself. This is a subtle but profound difference.

In the 1930s' Edwin Hubble made the remarkable observation that the universe is expanding, everything is becoming further apart in space from everything else. A standard metaphor used is that of an inflating balloon, each measurable point remaining in the same relative position yet becoming ever more distant from other points. The current big bang theory assumes a specific origin of this inflation, an imaginary air supply for the balloon.

Currently the smallest incremental measurements possible by humans are those of a Planck length. Consider this for a moment as the point of origin, the entry point of the balloons air supply. Also consider this as occurring at all possible points, including not only inside planets, but inside our bodies and physical brains themselves; *everything* is coming apart. There is no such thing *other than awareness* as a single point from which to measure that is not itself subject to this expansion.

If the universe is expanding then a source, or point of origin of space itself need apply. A problem in identifying this source is created by scale of human measurement, we make our observations from what we believe to be a stable platform, but the entire platform itself is continuously expanding.

Because our awareness is mostly the product of relationships between countless physical parts that are receiving information regarding other energies also expanding in space, any measurement made will be different a nanosecond later; we are forever playing catch up with reality, observing after the fact, physically incapable of apprehending the present moment. *The continuous, stop-action measurement of change creates this idea or perception that we call time; time is not a component of space, it is awareness of changes to energy within the venue that is expanding space.* The beginning of time as we know it exists at a physical location everywhere in and around us.

We readily acknowledge this on a large scale; we admit that light we are perceiving from distant stars "happened" a long time ago. Why we think this is any different for *all* sensory information at all scales is a puzzle. Perhaps it comes back to thinking of our physical bodies as being *what we are*, instead of timeless points of awareness.

A Different Frame

The point of origin may be smaller than a Planck length, but that's currently our limit of measurement. Possibly this defines our limit *because* this is the point of origin, time as we experience it literally begins at this infinitesimally tiny point, a point that exists everywhere.

At the other end of our scalar measurements we have the sensory boundary delineated by the speed of photons; to us the distant edge of the universe seems to be disappearing, rushing away from us into apparently unknowable space. This perception of physical distance is relative; if the expansion is occurring at every possible point, then our scale of large and small is completely arbitrary, awareness defining a center of the possible frame.

Human senses, amplified by technology or not, have narrow, fixed parameters with which to measure reality. Our window on the universe is defined by which senses we pay attention to, both individually and scientifically. Focusing on one sensory source of information acts to dramatically limit the scope of this window; we become quite literally blinded by the light.

A century ago Einstein eloquently demonstrated the equivalence of mass and energy. Everything we can possibly measure is comprised of energy interacting with itself in some form. Though the venue of space is expanding, energy is communicating with itself at different speeds, at varied distances, in countless fields. Because we ourselves are expanding along with the universe we cannot triangulate on reality itself, we can however measure some changes in energy relative to our singular position.

Energy appears to slow the expansion of space at a rate directly correlated to the density of the energy. General relativity* states that mass/energy warps space; imagine as if it is doing this by *slowing* the expansion. Experimentally it would present the same whether one thinks it is space or time that is being "warped". This might be why we would perceive mass to become infinite beyond the speed of light; it has moved outside of our timed measurement scale.

The window of human measurement is limited not only by the scope of our senses, but also by which scalar field and window within that field we choose to make observations within. The *relatively* high speed (smaller scale) at which sub atomic particles communicate makes it impossible to make usefully accurate predictions; the system has completely changed by the time the information can be presented to our awareness. Larger scale measurements are slow enough *relative to our scale* to make useful predictions; remove a boulder from the moon to reduce the lunar mass and we can still calculate an orbit.

If this idea is true, then neither the Copehagen* nor Hugh Everett's interpretation of quantum mechanics is correct. Everett was closer, but

A Different Frame

even his Many Worlds* theory is based upon single events, forming linear paths.

In the Copenhagen interpretation, believing that the wave function collapses when observed/measured is the ultimate of anthropomorphic silliness. The universe is one giant endless wave function or matrix, observation of any given event arbitrary. Conscious observation of an event is *always* after the fact; there is no such thing as a live broadcast of information for any single point of awareness. The thought experiment involving Schrodingers' cat* is merely a representation of a potential future measurement, an admission that we cannot ever perfectly predict the future.

The Many Worlds theory is wonderful for the imagination, but seemingly has no basis in reality. The overlap of all energy relationships is occurring in every possible field, direction, and scale simultaneously; it is physically impossible to measure this because as soon as we define a field and window of measurement everything has already changed. The belief that any single measured event is creating a whole separate reality or universe reflects our inherent narcissism. There is only one reality, and it is the ongoing summation of every energy event in the history of the universe, at the present moment. We are forever viewing snapshots of what it has been.

We have recently determined that there is a huge amount of energy in the universe that although we can see its' gravitational effect, we are unable to measure it directly; this has been labeled "dark energy". The currently most popular models posit it as acting as a repulsive force to gravity, accelerating the expansion of the universe.

What if there are types of energy that are *not inflating* within observable spacetime? Perhaps they are too dense/moving too fast to "squeeze through" a Planck length to go sailing into our view of time as we experience it. On the other side of the scalar screen where time stops there could be a whole vast universe of energy inflating into what we would consider the infinitely tiny throughout all of space. Inflation would then be occurring in an ever-smaller direction, *not just outward*, which would be a completely natural result if the inflation is happening everywhere; small and large becoming arbitrary frames solely dependent on a scale relative to human perspective. The results of the big bang are then manifesting at every possible point in our 3D space, not some imaginary fixed location in the distant past. This would create the very definition of another dimension, a seemingly timeless dimension that is here, everywhere. Time is not a fourth dimension to add onto our three dimensional universe; expanding space provides a dimension, format, or venue for an infinite number of directions to be possible from any given point. *The physical location of the*

A Different Frame

origination of expanding space forms the edge of this other dimension.

Contrary to acting as a repulsive force, dark energy could then be what is actually "creating" gravity. If we think of gravity as being a force that is attractive of energy to itself, and there is a large amount of non-inflating energy sitting on the other side of the Planck length boundary, this might account for gravitational force as we know it. Electromagnetism, and the strong and weak nuclear forces would be different modes of energy communicating with itself within expanding space, but the overall effect of energy attraction remains. This might be why Einstein couldn't reconcile gravity with general relativity at small scales; the energy requirements simply don't exist in measureable expanding space.

If we have masses of energy such as planets that are simultaneously slowing the outward expansion of space and adding to localized gravitational force, and there is a large amount of energy operating on the other side of the Planck length boundary, it there any reason to assume that it wouldn't be clumped as planets and people are in this dimension? Clumping of non-expanding dark energy could manifest in our measurable space as black holes, only instead of "trapping" energy, it would be an outpouring of it in all possible directions, a sort of rip in the Planck screen at the base of our inflating 3D dimension, dark energy *measurably present yet still not expanding*. This would nicely explain their powerful gravitational force, as well as their apparently timeless nature. A massive black hole in this dimension might appear as something like a large star on the other side, our stars appearing as black holes to an observer from the other dimension. The universe would seem to be bifurcated by time itself, an observer from either side of the timeless point experiencing time/inflation into vast space away from them, and disappearing down into the seemingly tiny point of a Planck length.

This type of model might also resolve many of the theoretical conflicts surrounding matter and anti-matter, primarily why there is very little anti-matter in our observable universe. Just like dark energy, if one thinks of expanding space as only a portion of the universe, and time as we know it exists in a fraction of that part, it is clear that there is a whole lot of energy/matter that we are physically unable to observe and hence measure.

Thus the point of origin of the big bang could be everywhere and everywhen, space expanding in all possible directions creating the perception of time. Energy is drawn to energy, creating gravity in all possible directions. Space expands, energy contracts; the universe is breathing. Time and biological awareness are synonyms.

The idea that there is a 4th dimension of time is misleading; there is only energy, awareness, and expanding space. As Freemen Dyson* put so succinctly; "it's infinite in all directions".

CONSCIOUSNESS

So we have two major players comprising the venue of reality: space and energy. There is a third field which is most relevant to all life, and that is awareness. This field of consciousness is truly the most baffling of all, as we use the field itself in our attempts to measure it. The metaphor of not being able to see the forest for the trees seems quite apt.

Space and energy are what we have been scientifically observing and therefore measuring all along. These appear to be "things" that are external to ourselves; we observe changes within a window of time and then we witness other changes at other points in spacetime and then draw conclusions as to what fundamental rules might be governing them. It is the repetition of these rules within a variety of contexts that allows us the ability to understand something of their scope and nature. To understand awareness from a scientific perspective we should apply the same model to consciousness, and look for the rules that seem to occur everywhere.

First it seems necessary to define awareness. As previously discussed, awareness is a biological single processing point that receives sensory information from outside of itself and makes decisions in response to that information. These are biologically identifiable points in space where data external to these points is received, processed, and acted upon. In this manner it is one of the fundamental components of life, even plants and microbes are aware. Since it currently seems difficult if not impossible to determine if a plant is *self*-aware beyond physical response, for our purposes here *self*-awareness may be considered consciousness. For this discussion we are addressing awareness, not self-consciousness.

It matters not if an external source of sensory information is itself alive or inanimate, or if it is perceived by other life forms, or even if it is present a moment after perception, reality is what it is to that awareness. Debating the validity of this point seems a narcissistic waste of time, a judgmental discussion that *Homo Sapiens* are quite fond of. Many humans feel a need to discount other beings reality if it doesn't satisfactorily match their own; by doing so they become blind to the forest. The universe then becomes one dimensional, a dimension limited to a single being's relationship with it. It seems incredibly sad that so many humans operate this way, their journey here and now being viewed through such a tiny window. At least this waking awareness part of the journey seems to be only a fraction of the experience of being.

If we view the expansion of space as creating a venue to witness energy and hence the perception of time, then individual points of

A Different Frame

awareness cannot be expanding, or at least not expanding at the same rate. It is this relative *difference* that enables the illusion of linear time, the only fixed platform being that of individual awareness. Sensory information takes time to be obtained, processed, and delivered to the cognitive stage; the presentation upon that stage being as close to now/reality as any being is capable of. Even though it is after the fact, and it takes time to think, nothing can possibly move faster for any individual.

Although the individual platform of observation is fixed relative to the expanding matrix of space and energy that is occurring from and in all possible directions, that platform could still be created from parts that are expanding. As previously discussed, any "thing" that can be compartmentalized within language/thought is the product of many communicating bits of energy that are moving at different rates relative to the expansion of space and each other. An identifiable whole that is greater than the sum of the parts is created, the whole at hand being that of awareness.

Awareness therefore operates within a field independent from space, yet is able to receive information from there via a wide array of species-specific sensory channels. Even though individual points of awareness may be expanding away from one another in space, there must be underlying forces/rules governing them all; replication of results stems from fundamental properties of the universe, and for points of awareness to be created, there needs to be a common mechanism at work. This organizational system that creates these localized sensory reception points would represent an entirely separate field or force within the universal structure. This platform of observation, or awareness, would add a third primary element to reality, expanding space and energy being the other two.

Modern western psychiatry "discovered" the subconscious primarily through the works of Sigmund Freud and Carl Jung. While there have been countless discussions regarding the theoretical differences between these doctor's approaches, there exists a primary division that is relevant here. Mr. Freud proposed a basic model comprised of the id, ego, and superego. However accurate or incorrect one may feel this to be, it was a model of a communicating system that played out in a unique fashion for each individual. According to Freud, these fundamental components would interact with each other based upon how they had been molded by the experience and subsequent cognition of the individual. This would form a base platform of the psyche, shaped by individual events and processing of those events.

Mr. Jung on the other hand, presented the concept of the collective unconscious, whereby information structuring the psyche was at least partly shaped by a *field* external to the individual, not just experience. This field contained what

A Different Frame

Jung called "archetypes", which existed independently and external to individual experience, yet were interpreted differently by each person.

A metaphor for these theoretical differences might be that Freud proposed that we were blank 3-part hard drives when conceived, and Jung would say that not only did we come with some software installed, but wireless capabilities to boot. If we were discussing computers they could both be equally correct, but we are talking about fundamental properties of the universe, not technology. It is quite clear that the base instructions necessary to produce living awareness are naturally occurring and universal. If we then understand that animals, insects, plants and microbes possess awareness to some degree, and that the physical hard drives are radically different from species to species, yet this thing called awareness is still present, then *the field must exist independent from human experience.* One would seriously doubt that a wasp's psychic structure would be able to be usefully interpreted by the relationship between its' id, ego, and superego, although it may be theoretically possible. It seems obvious that there is a greater external executive system or field at work.

So how to discern this greater field of awareness? Applying empirical methodology perhaps we should look for those things that seem to occur within it everywhere and everywhen, the common factors. We already have a working definition of awareness that applies to all life, but being unable to directly inquire as to what awareness is like for a houseplant or insect in human language, it seems reasonable to first gather information from the human realm.

There have been myriad expressions of human culture throughout our few thousand years of written language. We can make assumptions about what conscious experience was like for pre-historic *H. Sapiens*, but they left no direct statements of such. Archeological evidence of tools or dwellings is highly subject to interpretation, and arrangements of stone monoliths tell us little of the conscious experience of their builders. If written language is the tool that science and technological knowledge is being built upon, then perhaps the same should apply to the search for the field of awareness.

One common written factor expressed throughout human history, in every culture, at every time, at every place, is that of religion. We have not discovered a single large group that possessed the ability to write that did not document some form of a power greater than humans. Documented contact made with groups that had only oral communication revealed that they also maintained some system of religion outlining a higher power(s). The channels of communication that this power communicated itself to individuals and groups have varied widely, as has the form of the field itself. In some cultures it was intensely personal and direct, in others distant and vague. Sometimes the power had fragmented itself into multiple personalities, at other times it was unified. Sometimes it had conscious desires to manipulate the actions of our species, to deliver rewards and punishments as it saw fit; at other times it was detached, merely present and available for communication if a human were to initiate contact. In some religions it was concerned primarily with humans, in others it was expressed through all life equally. Sometimes it was singularly self-aware, sometimes not. The common documented message in every instance is that there exists a power or field greater than humanity, an organized force that is separate yet somehow communicates to and/or through us.

To define religion; a structured human set of rules that attempt to define the ways and means of this higher power. In this way, modern science is very much a religion; it is searching for the nature of this power, carefully documenting what it seems to consist of.

There are those who currently ascribe themselves to be pure atheists, believing only in their own reality or verifiable science itself; they are as blind as any religious zealot is to greater reality, paying attention *exclusively* to a fixed human set of rules that reality must adhere to. Unfortunately, many people have their sole connotations of a higher power tied to a particular form or forms of religion, and because they see ridiculously impossible statements made by religions other than theirs, they focus on the obvious fallacies and throw out the entire concept. It is like someone who hears another describe lightning as being hurled to Earth by an angry old man, and then goes on to write off lightning because this is clearly bullshit.

The only true atheist is one who believes that greater reality exists solely as a product of their own individual creation; the entire universe vanishes when they die. Scientists are the priests and gurus of this age, but the power is what the power is.

There have surely been many people in every society that did not internally subscribe to their culture's preferred religion. This in no way would negate the validity of a greater field of awareness; they are even using this field to make their assessment. When we ascribe a human desire to understand and/or control reality to be the *sole* root of religion, we throw out reality itself; we use ourselves to deny ourselves.

One possible way to view this force is as a universal field of awareness. Similar to an individual processing point of sensory information this field might represent a unified processing location or platform of *all* individuals' information. Any whole idea or thing we may conceive of is an imaginary box of communication between many separate parts; a grove of Aspen may be considered as individual trees, but they share a common root system.

A Different Frame

Whether this field of awareness is labeled God, collective consciousness, or some other word, it is clearly present and active regardless of the form of individual apprehension and interpretation.

So all of humanity has documented the experience of a field of awareness that is greater than the sum of individual humans. All too often there is a tendency to try to understand this field by studying individual human minds; this is like trying to create a topographic map by studying the engine of a specific type of automobile. Once we understand that there are many minds following common paths within certain limitations, we then need to look at how the entire group interacts with reality; how the whole communicates within the venue of space.

To return to physics for a moment; there is a large amount of energy that is not expanding to *observably* participate in the creation of our observable, or timed, dimension. The Planck length may be considered the point of origin, or edge, of this dimension, but reality does not disappear at this point, *it is merely unobservable within time as we perceive it.* Whether this seemingly timeless dimension is viewed as a contraction of energy creating gravity, an expansion of space into the infinitely tiny, or both makes no difference; points of awareness are present in expanding space to take in information.

These individual sensory reception points operate within a field that is independent of the expansion; the platform is fixed in relation to time. Yet the material components that create the localized individual points are subject to the expansion; our physical brains and sensory organs are literally coming apart along with everything else. In order to measure changes it is necessary to maintain a stable point of observation; if awareness was merely the sum of physical parts our perception of time would not be possible because we would be flowing along at the same rate as everything else; the whole of individual awareness is something different than the expanding mechanisms creating it. It is like awareness is a timeless passenger in a biological space ship.

In order to maintain the ability to measure the present moment there needs to be both a field of awareness and a means of localizing points within expanding space. If we imagine a scaled physical location to be where the separation of universal expansion and contraction occurs, or the edge of another dimension, then it would make sense that this location would be where the field of awareness operates. This locale would be the very definition of timelessness from at least a human scaled perspective, it is where dark energy resides.

If energy is gravitationally attracted to this location, and informational awareness is creating new arrangements of energy in the form of data within the field of awareness, then individual awareness would be subject to anything else that is gaining energy/mass at a specific location; it would be increasingly drawn towards a Planck length. If awareness operates within an arena that is relatively slower than the expansion of space, and the physical parts enabling individual connection to this field are expanding within space, there would appear to be a need to constantly re-calibrate individual position within the field. Some sort of process seems necessary to maintain a *central point of informational coherence* while simultaneously gaining energy and coming physically apart. To do so seems akin to trying to change the tires of a moving vehicle; a seemingly impossible feat, changing wheels only possible when the car is stopped.

In addition to a higher power, there is another universal factor to be found regarding awareness. It is so ubiquitous that that the vast majority never stop to consider its' function or purpose, and similar to religions, those that do ponder it have not arrived at any sort of consensus. This factor is sleep.

It is known that all terrestrial life forms experience circadian cycles; or changes to levels of awareness based on a 24 hour period, regardless of whether light is present or not. This has been demonstrated even in plants and microbes, the need to regularly close down awareness is as mandatory for life as water.

What a plants' feeling of sleep might be is difficult to ascertain, their experience of awareness being so alien to our own that it is almost impossible to discuss. Many botanical species exhibit obvious periods of dormancy, but many others externally seem to function in a state of constant activity. Because we have yet to identify a localized point of awareness for a plant, the idea of floral sleep as we know it is mostly intangible. Not so with fauna, here we can easily measure changes of conscious states and responsiveness; a localized point of awareness is able to be measured by both observable behavior and electrically within the brain. The relevant point is that sleep in some form is an essential component of individual points of awareness, defining it as clearly having a direct connection to the greater field of consciousness.

Perhaps sleep is the mechanism/process by which a point of awareness is re-oriented within expanding space, the *shutting off of time* necessary for calibration. It is impossible to shut off space, but time is individually experienced. Our individual storage systems are limited in capacity, and would quickly become gravitationally overwhelmed in memory if there is exclusively input, which would prevent new information from being acquired. A constant flow of input only would cripple the ability to discern time itself as an independent measurement tool. *This acquisition of timed sensory data is the very definition and function of waking awareness.*

To shut down the flow of new information seems absolutely necessary, not only to sort,

A Different Frame

consolidate, and dump data, but to literally return to the base of timelessness. For a point of awareness to maintain that relatively fixed platform of observation, it must *connect* to that platform somewhere. In order to exist independently of both gravity and expansion, regular returns to the beginning of time and connection to the field are needed, a reset of the clock. Without doing so, informational coherence is lost, the individual mind becomes nothing more than energy existing in space, no longer able to act as a sensor for the field. If the platform is timeless, it cannot consist of information structured by time. Individual waking awareness is like a tree that is able to separate itself from it's root structure and go wandering over a hill to see what's there. It would perceive things it's fellows could not, but without returning to its' roots it would eventually cease being an identifiable tree.

The field of awareness is required for the individual points to exist, and the field itself consists of the points. If sleep did not occur there could not be individuation within spacetime, we would become like computers, merely timed measurable points of mechanistic probabilities expanding in tandem with the universe. There could be no platform of observation from which to perceive the ongoing coincidence of simultaneous reality, there would be only space and energy. Awareness of time could not exist in an expanding universe without regular closures of individual awareness.

The issue of a 24 hour sleep pattern across all terrestrial life is notable not only for its' necessary presence, but also for being identifiable within a 24 hour period; why a day? The obvious temptation is to conclude that it is based on light, as there are measurable periods of its' presence and absence in any given day, but there are valid reasons to discount light as the sole mechanism of circadian cycles.

The first reason is that there are many life forms that have evolved and live in complete darkness and still exhibit 24 hour cycles. Deep undersea and far beneath in caves, life still sleeps within a 24 hour format. A human blind from birth will require sleep in time frames similar to sighted people. Many life forms adapt well to peak conscious functioning independent of light availability, nocturnal or diurnal tendencies are able to be individually adjusted.

The other consideration is that of hibernation. Life may be maintained for much longer than 24 hour periods if it is asleep or comatose, but waking awareness requires breaks. It seems clear that there is another factor at work in determining the window of regular timed shut downs of awareness.

What about gravity? This attraction of energy to itself is present always, completely free of the effects of light and temperature. We can now measure the effect of gravity *on* such things, but these variations are not changing the fundamental nature of the force. Gravity bends light and traps energy into identifiable mass, the principle of attraction is all encompassing.

The most gravitationally powerful masses most immediately effecting terrestrial life are the sun and Earth. Even though gravity is attractive everywhere, in every possible direction down to a Planck length, as energy forms into mass it exerts exponentially greater attractive force around itself. Because the Earth rotates relative to the sun, this changes the amount of gravitational force at any given Earthly surface point on a 24 hour cycle. If each point of awareness is on a fixed point on the globe, the degree that the sun/Earth alignment is effecting gravity at that point is continually changing in 24 hour cycles in relation to the gravity of the sun. These changes would not only be caused by the every-changing distance from the sun, but by alignment of the mass of Earth in relation to any surface point and the sun.

If the function of sleep is to shut down individual awareness and hence perceived time, this might be achieved by a gravitationally driven return to the timeless field of awareness. To *fall* asleep would allow for the return of conscious energy to the place necessary to maintain a relatively fixed point in expanding space, say just smaller than a Plank length. A possible metaphor for this might be a bouncing ball; the ball needs to fall back to a solid base in order begin its' upward journey. If waking awareness is a single timeless point traveling through expanding space it must return to the source regularly to localize and individuate itself. Otherwise it becomes just another bit of energy riding the expansion, unable to calibrate it's own relative position within the universal matrix. In this way, each point of awareness is a small part of a universal singularity poking into expanding 3 dimensional reality.

This bi-dimensional model of physical awareness would provide a platform to explore all of what are considered "fringe" areas of cognitive science. A timeless unified field of information that contains the sum of all awareness would represent the ultimate source of otherwise inexplicable psychological phenomena. Remote viewing, precognition, near death experience, synchronicity, and many more seemingly inexplicable manifestations of awareness would have a common root in this second dimension; a dimension seemingly very small, but actually massive by being present everywhere. Time travel, at least of informational awareness, becomes a very real possibility.

A POSSIBILITY

For those that survive the next major global war, i.e. one that spills the full might of superpowers outside of the Middle East, there are solutions that could potentially avoid *homo sapiens* demise by unintentional suicide. It is likely an unfortunate fact that global catastrophe is necessary to humble us enough to try different methods of governance and finance. Hopefully the horror will be primarily financial, but this seems unlikely. George Orwell was probably correct when he stated that in relation to governments currently in power, the perennial number one priority is to remain there.

It seems impossible to sustain human society without some form of government. There are those who promote anarchy, but they are few, and quickly fall to fighting amongst themselves over who does it the best. An organizational structure that holds and promotes the long term well being of humans as a species is seemingly required; individual nations as they currently stand will ultimately destroy us. There will always be those who wish to take more of everything at the expense of others; acknowledgement of this incontrovertible fact may allow us to incorporate greed instead of vacillating between embracing and loathing it. Our current economic standards of capitalism and communism are perfect examples of this polarized scale, but a third option that utilizes the best of both of these is now technologically possible.

Utopian society exists in the mind of the beholder; there will always be some dissatisfaction with governmental structure from any individual perspective. Since administrative perfection is then eternally unachievable, the realistic goal then becomes one of governmental stability and sustainability. Up to the present day, and perhaps until the death of our species, we have been unable to create either of these conditions for very long. The ongoing polarization of wealth and corresponding warfare inevitably destroys nations, and if left unchecked will very likely be the end of us. If we are to develop a realistic means of governance that provides both stability and sustainability in the long run it seems we must address this process of polarization that rips us apart.

The framers of the US constitution understood the fundamental problem well; that government will seek power for its' own ends over the greater good of the constituency. The system of checks and balances that they built into the structure of the nation was designed with the specific goal of preventing accumulation of power in any one arena. At the time it was possibly the closest humans had yet come to a structured democratic ideal, but change has been occurring very rapidly this past 200 years, and is only happening faster with technological advancements.

The prime enemy of democracy is partisanship; always has been, always will be. A primary component of democracy is equal representation within group decision making; one person/vote is as valid as the next regardless of gender, race, money, religion, etc. One reason this can never exist in perfection is that there will always be powerful and wealthy groups that believe and act as if they are in some way superior and therefore better able to make the correct choices.

A small business as well as a corporation both have the same primary goal; to make as much money as they need to pay the bills, plus have some left over. Unlike private businesses, governments can borrow money from citizens who haven't even been born yet, but they still are beholden to budgets and balance sheets. Every political and business decision is inevitably dependent on finance, and with money as the base priority the people with the dollars ultimately make the rules. Money represents the ultimate partisanship, and thus the greatest danger to democracy.

Another reason that pure democracy cannot exist is that it is physically impossible for every person to vote on every decision. Representation seems necessary, and once power is given an individual, all the usual forces of money, power, and special interests hold sway. Even the most honest and high-integrity representative must compromise their best effort at times, if for no other reason that there are limited dollars. Corruption takes many forms, both subtle and obvious, but the leverage is usually exercised through individuals.

Reality is the only pure democracy, all events are equally necessary to produce the moment. Human political systems are far removed from this by basic physics, but the rules we choose to make and adhere to sometimes increase the distance exponentially.

As previously discussed, this is a good thing from an evolutionary perspective, mutations of the group are necessary for ongoing adaptive responses. Unfortunately when money is the ultimate deciding factor our systems become self-consuming, severely restricted in both exploitation and exploration outside of humanities' little box of rules. We possess the resources and technology to provide the basics of food, clothing, healthcare and shelter to every human on Earth, and yet we cannot. Money, our own self-designed communicative tool, defeats us by its' imaginary limits.

Surely somewhere there is an ET shaking its' heads in disgust, seeing all that humanity is technologically capable of, yet unable to achieve so much because we can't find a way to pay for it. It would be laughable if not so sad.

Money

A working definition of money; a medium of exchange represented by coins and bills, as well as contractual banknotes. It is a written language whose connotations are instantly translatable to material goods and services, it is the language of any large scale human cooperative endeavor. Coupled with mathematics it is the global language of humanity, and like all written languages it comes in many different dialects, but serves the same function regardless of the shape of the mutually agreed upon symbols.

If written language is enabled by the cross-referencing unified code between visual and other sensory channels, money adds a third dimension to human language, that of physical presence. Unlike ideas and concepts, money is something that is concretely measurable to be there or not. It creates a grounding of meaning to the physical world that permits an entirely new format of intra-species communication. Whether paper currency or digital records, money translates exclusively in the physical realm.

The idea of money has become so ingrained into our consciousness that we find it difficult to imagine a world without it; like a fish trying to imagine a world without water. But it really only influences human interactions; this is because, like any other human language, its communicative power is derived solely from mutual agreement.

It is important to consider that money has been a relatively recent invention on an evolutionary scale. A few thousand years may seem like a long time, but compared to how long humans have been functioning in groups on Earth it's a drop in the bucket. The language itself has evolved tremendously; picture a Roman soldier trying to envision global computerized stock exchanges.

So humanities greatest inventions, the ones that define us in a unique place on Earth, are written language, the technology it enables, and the physically grounded language of money. Both written language and money are generated and maintained exclusively by mutual human agreement and thus exist independently from ongoing universal events. Neither writing nor money can actually change anything in physical reality within the present moment; one can write lies and it doesn't change what really happened, and all the money in the world can't do anything by itself. They are purely human generated and maintained communicative tools that facilitate cooperation as a species.

With money as the priority the principle of altruism is rendered impossible at large scales; an individual person or small group may intentionally sacrifice it's own life for the survival of others, but a human organization, especially a government, is incapable of this. Indeed, they must continue to exist in order to fulfill their organizational purpose. Even the most well intentioned non-profit charity organization must prioritize its' own life over others; this is because it is dependent on money to communicate *as* an organization with the rest of human culture; any human organization that exists within the context of a greater system must have money to survive. A small group of humans may be able to independently survive by living off the land in a remote location, but if a group is to be considered a factor in the broader evolutionary relationships currently in play on Earth money is *the* means of survival.

Because money's communicative power rests exclusively on collective human rules, it is one of the only things humans have any real control over. As long as it can be discussed with the language of mathematics the possible forms of transactions are almost endless.

Money as a tool has been evolving since its' inception; like any language, changes in form necessarily occur to meet changing conditions. There have been a couple of identifiable major shifts in attitudes towards finance that are important to consider.

The first coins were minted in Turkey somewhere around 600 BC. For the next two thousand years of its' existence money had value backed exclusively by being made of precious metal, and it carried that value with it regardless of who shaped it. Possession of the physical material itself was all that was required, and securing sources of metal, usually gold, formed the very foundation of economic growth.

The physical metal itself *was* money, no psychological belief system beyond its' immediate presence was needed.

The first major evolutionary step was made in 17th century Europe with the invention of banknotes. These originated as receipts for deposits of gold, and grew to reflect that they could be payable to the bearer of the note, not just the depositor. No longer was carrying the metal itself necessary, far more value could be easily contained on a piece of paper. It is no small issue that this change required *trust* in the bank that the gold backing the paper was actually present, and for the next few hundred years it was a frequent occurrence that too many people tried to redeem their cash at once and found it wasn't really all there. Still, the psychological shift to placing *faith in an institution* as opposed to personal possession of the metal itself is significant. It was the belief that the bank really had the gold that lent relevance to money; the bank might really have nothing but if people believed it did, paper money worked just fine. This cognitive leap of faith perfectly set the stage for the next evolutionary turn.

Fiat currencies are the standard to the present day. These are monies not backed by any single specific bank, but by the promise of a national government to pay. Up until 1971 a nation still needed to maintain substantial gold on hand to back up this promise, but no more. There is now not a single national economy whose money is based on anything other than a promise. The

A Different Frame

value of money is now exclusively dependent upon the *perceived* ability of a government to pay its' debts. The significance of this change cannot be understated. Money as a language of physical presence has moved into a purely psychological and mathematical realm; a realm controllable exclusively by globally powerful national governments.

The most recent evolutionary iteration of money is cryptocurrency. Our trust in value has evolved from a physical metal to a private institution backed by possession of the metal, to governments at least partially backed by metal, to purely belief in national government. To take this the next step in trusting computational algorithms could enable us to take an evolutionary leap forward comparable to abandonment of the gold standard.

Blockchain technology was originally introduced with the intended function of creating and growing money. This is only one possible use of this revolutionary computerized process; the applications range far and wide, but it is the generation and distribution of money that concerns us here.

Cryptocurrencies are generated and maintained exclusively by computer algorithms, a true cybernetic loop. They exist regardless of physical or sociological resources, needing no physical backing of any sort to be communicatively valuable. The value is created purely by the digital interactions between computers, wealth manifesting from digital air. Fiat currencies still ultimately require mathematical connection to the physical resources of nations if they are to retain value; not so with crypto. Never before in history has a monetary system had the potential to be completely free from human manipulation.

Much of the current debate surrounding cryptocurrency centers upon its' seemingly uncontrollable nature; governments and individuals alike desire to exert control over financial generation and flow in order to communicate within human culture. The ability to regulate who has access to money is the very foundation of financial value; if dollars were scattered around the globe for everyone to possess all they wanted they would be worthless by the simple fact that anyone and everyone had them; it is the *desire* to possess and control money that lends to value to it. Just as written language, money must be limited to identifiable, mathematically accountable increments in order to be useful.

The founder(s) of Bitcoin, the first cryptocurrency, established the base release of ever-increasing increments of money to be obtained by computer maintenance of the blockchain; wealth is generated by raw computational processing. Thus, those that control vast arrays of computers performing "mining" operations ultimately make more money. The more computing power any individual or group controls, the wealthier they may become; the foundation of growth is based purely on physical computer resources. The fact that these computers can be had by anyone, anywhere in the world is a completely different paradigm from nationalized fiat currencies. Gold mines had specific geographic locations that could be controlled, and a nations' borders can be physically enforced; the value of crypto resides solely in cyberspace.

It is important to consider that the value is still a function of psychological trust. That this trust is placed in computers instead of human administered government defines the evolutionary step.

A problem inherent in blockchain technology is one derived from its' strength; once a transaction is recorded, it can't be reversed. Any binary digital system that is open to human input will ultimately be hacked in some way. Because the nature of a blockchain transaction is forever fixed and essentially untraceable, successful human manipulation of the system would be the perfect crime; one never revealed in consensual reality.

Another major weakness of existent cryptocurrency is that having competing forms allows for value manipulation. Controlling large sums of multiple cryptocurrencies and computing power allows for dumping of one and/or buying another, directly and immediately affecting the value, just as is commonly done with fiat money. If one party controls enough computer resources dedicated to crypto mining as well as large sums of the currency itself a shell game becomes possible, foreknowledge of shifts in resources being the essential ingredient, the perfect insider trading. With fiat currencies this manipulation is able to eventually be compensated for within the greater system because the base value of a company or nation is ultimately derived from real-world resources; the company or state has what it physically has independent from perceived monetary value. Not so with crypto; it's value lies exclusively in the cyber realm, and so can be directly affected by any single person or group who controls and mines enough of it.

So it seems that the only way cryptocurrency can be truly stable and sustainable in the long run is for there to be only one, and that one to be maintained *exclusively* by a single *closed* cybernetic AI program. Until this happens, the broader cryptocurrency market will continue to be merely an imaginary computerized money game, nothing more than a digital casino owned by an anonymous oligarchy. We could do better.

If humans are to continue to have both money and government, it seems another major evolution is necessary to slow financial polarization on a global level. As previously discussed, we can ill afford many modern world-wide wars. To be able to have our invention of money serve our entire species *and* the people who have money might seem a pipe dream, but there is a possibility.

A Different Frame

A future possibility that may allow for long-term stability of our species is to have a single AI moderated global cryptocurrency whose sole application would be the financing of government. Nations, and thus private sectors could still exist in the same financial format with nationalized fiat currencies; a single global money could be an addition to, rather than a replacement of, existent systems. A single stable currency equally valued by all national governments could change the very definition of a nation, reincarnating humanity into an entirely new cooperative entity, an entity that would spend far less energy exploiting itself, and far more energy exploring. It would utilize our invention of money for everyone, everywhere, giving ourselves the gift of global organizational stability necessary for long term survival.

The European Union has been a recent large-scale experiment with a new type of fiscal unity, but because the Euro is just a larger scale fiat currency it cannot achieve the sort of stability necessary to avoid global war or radically accelerate innovation. However, by unifying multiple nations with a common currency it does help prevent war between those nations. A single global monetary base might do the same thing for our planet. Imagine doing something like the creation of the Euro for the entire world, except each nation would keep their own currency for individual and corporate wealth.

Many of the problems that have arisen in establishing and maintaining the Euro as money have resulted because it is used for everything, public & private alike, and individual nations have radically different economic capabilities. We have yet to try having a currency that exists purely for government.

As the system currently stands there are vast unknowns explored by economists, who, like all scientists, are able to produce probabilities. The exploration made possible by these unknown financial variables is what allows for mutations within an adaptive system; competition for money breeds both quality and innovation. This is a basic necessary function of technological evolution, and the private sector can continue to fill this need. However, it is equally obvious that stable government must exist as a platform for this process to take place.

Viewed from a contemporary fiscal perspective, governments and businesses are barely distinguishable from one another. They both are fundamentally dependent on their sources of income; consumers for business, business and citizens for government. They each are in a symbiotic fiscal dance with their resources; if either party is depleted or dies, so does the other.

It is not only acceptable, but desirable, that some businesses fail or are assimilated; this is the essence of evolutionary strength, and the core of capitalism. Even though nations fail, this inevitably involves the exploitation of war, which can now result in an abrupt end for our entire species. Even if we don't completely wipe ourselves out in WWIII, getting knocked back to the stone age will slow our technological development considerably, retarding humanities' adaptive responses. Stability of government is directly linked to the odds of our survival.

The important financial difference between the private sector and government is that of priority; governmental bodies' fiscal purpose is to regulate and distribute money with an eye to the long term future, while the main priority of business is to make money today. Ultimately financial decisions are based on a timeline; short term involves higher risk with potentially greater losses or returns, longer term is safer but with lower potential yields. If a business is offered some cash today vs a larger amount paid out over years accruing low interest it is completely clear that the business should take the money today; it is then able to invest the money itself for it's own ends and potential greater returns. Governments however, must play the long game if they are to survive.

When financial decisions are made by humans it is inevitable that partisanship will play a role, even if the partisans consist of an entire generation of humans. Massive debts are accrued in order to pay for what is wanted today, increasing the financial burden and risk for tomorrow. Constant predictable growth is required to maintain such a fiscal shell game, and there is no such thing as perfectly predictable in a human run system. Sooner or later a black swan appears, and there is a significant disappearance of wealth, vaporizing large portions of it almost instantaneously. Closed AI management of governmental wealth generation would radically reduce associated risks.

Cryptocurrency and AI are made for each other. The fixed nature of money can be completely maintained within a computer system; a 1 is only a 1, having no other components necessary to be what it is. The extraordinarily complex relationships formed by the basic binary format of credit and debit tied to accrual and depreciation rates over time are all able to be tracked by AI. AI can only be partisan to itself and fulfillment of it's role, it does not waste time arguing with itself about possible future financial projections, it is able to see what is possible or not within the language of mathematics. This seems to be as close to financial stability as is possible.

Removal of the profit motive from government would have incredibly beneficial effects for our species. Overall it would reduce both the frequency and severity of the polarization of wealth so closely correlated with war. It would significantly reduce corruption within the system, allowing legislators to make decisions that are more aligned with the good of their constituency instead of corporate interests. The absence of federal taxation would radically boost the financial strength of the private sector, putting more fiat currency in the hands of private business, local governments, and consumers. Most notably the presence of a consistently stable platform would increase risk taking in the private sector, accelerating the rate of innovation and hence adaptability of our species.

While the European Union is the recent historical example of monetary unification, there are many variations of what this could look like in the future. Here are just a few of the possibilities.

Have one cryptocurrency that is exclusively maintained by AI whose sole utility is paying federal expenses. This electronic administrative currency could be the base with which all fiat currencies are pegged/valued, much as the US dollar has been for the past several decades. What has made the dollar so attractive is it's stability, but this is bound to wobble and tip as ever-expanding globalization takes place. Not to mention that having a single nation having control of the global currency value adds to global polarization and hence war.

The initial valuation of all global national governments would be a difficult number to arrive at, but a necessary one to begin the process with. This number would be the base from which AI could launch a fixed percentage rate of growth, which then would allow for stable and informed budgetary decisions. Perhaps smaller national economies may elect to give up their own monetary system to be subsumed into a larger nations fiat currency, similar to the European process. It is difficult to envision how this could happen in a slow, moderated fashion, which is why global financial catastrophe will likely be a necessary precursor to such a system.

Borrowing may still occur within a global governmental cryptocurrency, but only within AI moderated mathematical realism, perhaps nothing more than a decade or two out, and limited by real ongoing needs. Putting future generations trillions of dollars in debt before they're even born would be seen as the pure narcissistic indentured slavery that it is.

Obviously a big question is what federal expenses specifically would be. Most likely they would be the minimum to maintain a stable platform of society; security & rescue, utilities, judicial systems, basic healthcare and core food supply. Fixing budget line items whose allocations float in direct proportion to both utilization and generation of governmental crypto wealth would be necessary.

One of the more ridiculous things about our current system is that a government agency must spend its' entire annual budget allocation if it hopes to receive the same amount of money the next year. There is no incentive for efficiency, only to spend and grow for the sake of growth. AI tracking of daily utilization of government services would provide for real time assessment and allocation of money based on actual needs and production, not partisan interests. Countless federal jobs would be

A Different Frame

created and destroyed based on real world needs in real time; the era of governmental laziness, apathy, and greed would end. It is no small wonder that such a system would be strongly resisted by many existent governments. Fortunately the private sector economy would be there, just as it's always been. This idea is a whole new *addition* to our existent system, not a replacement; there would still be poor people and rich people, just a lot less war & more innovation.

Competing cryptocurrencies could not be allowed in such a system; competition is the very root of innovation, but the enemy of stability. A single AI regulated global cryptocurrency coupled with existent private sector national fiat currencies would allow humanity the best of both security and risk.

A possible variation of this might involve the creation of individual regulated crypto accounts for each person, initiated at birth and terminated at death, maintained and administratively distributed exclusively through AI. This would fit nicely with biometrically validated identity tags, which are inevitable regardless of financial systems. The simple physical existence of each person would contribute their global "taxes" on top of the wealth already present in the system. The individual would have no personal access to this "account", it would be purely to pay for government budgets. This would also facilitate ongoing real time assessment of budgetary needs based on both location and usage.

There would be no dissention on who is qualified to receive a basic minimum umbrella of food, shelter, healthcare, and rule of law; it seems to be as close to a global utopia as is humanly possible. It would provide the best of communism in that the government "owns" everything of itself without interference of partisan interest or individual greed. A fully functional private sector with free markets based upon fiat currencies would allow for the best of competitive capitalism, with the security of a stable monetary platform to ultimately back them all. Surely we owe it to ourselves to leave the world a better place than we found it, and by doing so, give our descendents a fighting chance for a long future.

CLOSURE

Author Neil Donald Walsch presented an interesting idea in his series "Conversations with God". Actually he made many remarkable points, but one in particular stood out for me as a simple definition of God. I don't recall exactly where he said it, and I'm very much paraphrasing, but here's my recollection of that definition.

If you are aware of everything that has existed, exists now, and will exist in the future, there is no way to experience yourself. Experience is only to be had by interaction with something outside of Ones' own awareness; there must be an other. So God has fragmented Itself into an infinite number of points of awareness, each being essentially Itself, yet individually deprived of knowledge of the whole. In this manner God is able to experience existence. Life interacting with both other life and the universe is literally God having conversations with Itself, discovering what It is.

Understanding that we are all connected to this timeless platform of awareness allows for scientific methods to grow in scope. Appreciation of the different forms these connections take reveals the infinite beauty of life. Without those differences in form nothing could make sense, and without the platform nothing could even be perceived. Thank you.

GLOSSARY

Blockchain

From computer science/mathematics: A blockchain is an electronic digital list of historical transactions among multiple independent users that is secure and unchangeable. Its significance is that it creates a distributed ledger of information, thereby enabling a degree of security that makes peer-to-peer networks both feasible and potentially revolutionary.

Chaos Theory

A branch of mathematics, Chaos Theory posits that what seems to be chaotic or random is actually a causal system that is extraordinarily complex and highly sensitive to initial conditions. It develops a type of math that deals with dynamic, non-linear systems so as to identify cause and effect. The common illustration is the analogy of a butterfly flapping its wings in Brazil and a hurricane resulting in the South Pacific. The theory was summarized by one of its' primary founders, American mathematician Edward Lorenz, as follows; *"When the present determines the future, but the approximate present does not approximately determine the future"*.

Heisenberg Uncertainty Principle

From atomic physics, specifically German physicist Werner Heisenberg in 1927. The Principle states that the act of measurement itself alters what is being measured; e.g., the act of measuring the position of a particle affects its speed, and vice versa. If this is so, then the act of measuring can be neither passive nor objective. Also indicates that it is impossible to accurately measure both position *and* speed of any given particle at any given time. One of the foundational ideas of quantum mechanics.

Darwinian Evolution

A theory developed by Charles Darwin set forth in his book *Origin of Species,* published in 1859. This theory proposes a systematic process of physical change in living organisms by relation to their environment through time. It suggests that those individuals best adapted to their current environment will be those survive to reproduce, hence passing on their genetic structure along with the corresponding mutations that are best suited for survival. From a large scale perspective this process is similar to an individual life forms' interaction with the world; survival dependent on the ability to adapt to changing circumstances facilitated by some sort of change in which the form itself relates to the environment.

Will To Order

A seemingly congenital human psychological need for structure and linear hierarchies. Originally discussed by Aristotle, expounded upon by Aldous Huxley as well as other philosophers. Apparently the Will to Order forms a conscious platform necessary for comprehension of incoming sensory data, similar to an operating system of a computer.

Stephen Hawking

British physicist, cosmologist, and author, 1942-2018. Hawking was diagnosed at age 21 with a degenerative motor condition later identified as ALS, otherwise known as Lou Gehrigs' Disease. During and shortly after his formal education at Oxford he gradually lost almost all physical capabilities, including the capacities of speech, ambulation, and writing. As he lost the ability to communicate he developed cognitive methods of visualizing equations in geometrical formats that likely facilitated his breakthrough works in theoretical physics. Increasingly specialized technology granted him the ability to speak and write through a machine. Hawking was most noted for his work in discovering and describing black holes, as well as writing popular books discussing the cutting edge of physics in laymen's' terms.

The Borg

A fictional alien species from *Star Trek* that utilized artificial implants to create a collective hive-type mind, as well as adding some of each new species genetic structure to their own. They are perhaps summarized best by an address they give to the crew of the *Enterprise*: *"We are the Borg. Lower your shields and surrender your ships. We will add your biological and technological distinctiveness to our own. Your culture will adapt to service us. Resistance is futile."* The last three words of that quote have become often used in popular culture.

Cognitive Dissonance

From psychology: Mental discomfort resulting from simultaneously attempting to hold contradictory ideas, beliefs, or values. Classically viewed as resulting from personal internal contradictions as opposed to group relationships.

CRISPR

Acronym from genetic science: **C**lustered **R**egularly **I**nterspaced **S**hort **P**alindromic

Repeats. A genetic bacterial defense system that provides a platform for genome engineering. There are many systems developed from this base that may be utilized to edit DNA at precise locations; CRISPR-Cas9 is the most well known, and CRISPR-Cas13 has been developed for work specifically with RNA, and there are more platforms being developed on an ongoing basis. Because of its' ability to both model and edit genetic structure, CRISPR is widely promoted in medicine as a highly effective research tool, and more than 40,000 components of it have been shared with laboratories around the world.

Singularity

There are two primary definitions of a singularity. The first is a technological singularity, where artificial electronic computational power results in a self-replicating super-intelligence that causes unimaginable changes to human society. The second definition is used in physics to describe a point in spacetime where functional information becomes infinite and therefore immeasurable and unknowable.

Panspermia

A hypothesis that life travels throughout the universe by natural means, primarily via comet and meteoroid collisions with planets. It includes the possibility of intentional introduction of life by intelligent design, as well as unintentional contamination from spacecraft. Panspermia does not propose or address the origination of all life, it simply hypothesizes that it is capable of interplanetary travel.

Tesla and Edison

During the early years of developing commercial electrical power (late 1880's) Nikola Tesla and Thomas Edison were two of the major players in scientific research into how to best utilize this energy source. Both scientists were funded and otherwise supported by wealthy individuals, most notably J.P. Morgan, among several others. Teslas' research was focused on utilizing the planet itself as a source of energy, able to be accessed anywhere on Earth by building a network of towers around the globe that would provide a field so electrical power could be provided via portable units accessing this field. Edisons' company was initially involved with the research and development of direct current, but in 1892 was acquired by George Westinghouse, who had begun building a system based on alternating current (originally developed by Tesla) delivered from localized generative sources via physical wires to the point of use. This new company was General Electric, and they made good use of Edisons' research in lightbulbs. As J.P. Morgan and others held major stakes in mining, banking, telegraph networks, and General Electric, it was clear that the local control, and therefore profit, lay in alternating current distributed through wires. Tesla was cut off from funding sources and publically discredited as General Electric built a wired empire.

Extremophile

A usually microbial life form that survives and thrives in what are normally considered to be uninhabitable environments for life. Extremophiles have been found terrestrially in every naturally occurring temperature range, chemical setting, and radioactive environment.

Douglas Adams

British author, 1952-2001. Adams was best known for the *Hitchhikers' Guide To The Galaxy* series of books, and was a strong advocate for environmentalism and conservation.

Fermi Paradox

From Italian-American physicist Enrico Fermi, who in 1942 built the first nuclear reactor and is credited with being instrumental in ushering in the atomic age. Fermi famously pointed out the contradiction between the lack of evidence for extra terrestrial intelligence and equations indicating a high probability of it. Best summarized by the question *"Where is everybody?"*

General Theory of Relativity

From a paper published by Albert Einstein in 1915. Basically describes gravity as being a geometric property of space. To date General Relativity is the simplest experimentally confirmed theory of gravitational force, although it is unable to be reconciled with the sub atomic realm addressed by quantum mechanics.

Copenhagen Interpretation

From physics, specifically quantum mechanics. Developed in the mid 1920's, this commonly held model posits that physical reality does not have definite properties until it is measured. Because this essentially embraces Werner Heisenbergs' Uncertainty Principle, the Copenhagen Interpretation allows for generation of probabilities of measurement, and once direct observation takes place the "wave function" of reality collapses.

Many Worlds Theory

Hugh Everett's 1957 interpretation of quantum mechanics that posits the wave function of reality as endlessly dividing as opposed to collapsing. Each event division then creates/exists in it's own universe or dimension overlapping in time but not actual events. If this theory is true, there are a vast number of "yous" in existence right now, living their lives in countless other universes.

Schrodingers' Cat

From quantum mechanics: a thought experiment proposed by Austrian physicist Erwin Schrodinger in 1935. A cat in placed in a closed box with a flask of poison which is broken open after a random amount of time determined by radioactive decay. Highly reflective of paradox in quantum mechanics, until measured the cat is simultaneously alive *and* dead, only by opening the box and observing the cat directly does it become alive or dead.

Freeman Dyson

English-American theoretical physicist and mathematician, 1923-present. Mr. Dyson has been at the Institute for Advanced Study in Princeton for over sixty years, and has authored numerous books and articles on an astonishing array of subject matter.

www.ingramcontent.com/pod-product-compliance
Lightning Source LLC
Chambersburg PA
CBHW071029240526
45469CB00006BD/2144